Common-sense vs Evolution

Can Common-sense take down the idea of Evolution? It can.
Can Evolution stand up against a few common-sense questions? It cannot.

SHANE ANDERSON

© Copyright <<2022>> Shane Anderson

All rights reserved. No part of this publication may be reproduced, distributed, or transmitted in any form or by any means, including photocopying, recording, or other electronic or mechanical methods, without the prior written permission of the publisher, except in the case of brief quotations embodied in critical reviews and certain other noncommercial uses permitted by copyright law.

Although the author and publisher have made every effort to ensure that the information in this book was correct at press time, the author and publisher do not assume and hereby disclaim any liability to any party for any loss, damage, or disruption caused by errors or omissions, whether such errors or omissions result from negligence, accident, or any other cause.

Adherence to all applicable laws and regulations, including international, federal, state, and local governing professional licensing, business practices, advertising, and all other aspects of doing business in the US, Canada, or any other jurisdiction is the sole responsibility of the reader and consumer.

Neither the author nor the publisher assumes any responsibility or liability whatsoever on behalf of the consumer or reader of this material. Any per-ceived slight of any individual or organization is purely unintentional.

The resources in this book are provided for informational purposes only and should not be used to replace the specialized training and professional judgment of a health care or mental health care professional.

Neither the author nor the publisher can be held responsible for the use of the information provided within this book. Please always consult a trained professional before making any decision regarding treatment of yourself or others.

For more information, email commonsense_vs@outlook.com

DEDICATION

This book is dedicated to absolutely everyone who has ever wondered about evolution, how life began, or asked themselves, why am I here?

Note to You

The idea of evolution can be a very hot topic, a third rail some might say. Regardless of your thoughts on evolution, I believe this book will raise a question or two that you have not thought about or possibly vindicate a question that you might have asked yourself and wondered if you are the only one asking that question. Hopefully, you will at the very least learn something new reading this book, as I did in my research while writing this book. But most importantly, I want to challenge you never to stop asking common-sense questions. I hope you enjoy reading this book as much as I enjoyed writing it.

TABLE OF CONTENTS

Chapter 1: The Theory Of Evolution --------------------7

Chapter 2: Time Zero ---------------------------------- 11

Chapter 3: The Universe ------------------------------ 15

Chapter 4: Our Solar System--------------------------- 26

Chapter 5: Probability Of Life (Amino Acids & Proteins) -- 31

Chapter 6: DNA And RNA ----------------------------- 43

Chapter 7: The Single Cell ----------------------------- 51

Chapter 8: Microorganisms --------------------------- 58

Chapter 9: Great Oxidation Event -------------------- 66

Chapter 10: Snowball Earth --------------------------- 76

Chapter 11: Multicellular Life ------------------------- 89

Chapter 12: Eons---100

Chapter 13: Genus; Homo ----------------------------127

Chapter 14: Evolution Defeated ----------------------144

Full Disclosure ---155

My Faith --155

INTRODUCTION

Evolution: is it real?

I am standing outside early on this summer morning, sipping a cup of coffee here in the sandhills of North Carolina. I am looking at ants scurrying along a trail that makes its way from an Oak tree through the green grass passing in front of me where I am standing on a sand and gravel camping spot. These ants are marching along in single file, bumping into the other ants, their little antennas touching for a second, and then the ants continue along the trail, never venturing off to the left or right. Invisible to us but well-marked to the ants. Looking along the trail, I see the morning dew on little spider webs in the grass and some weeds growing among the grass. I wonder what those spiders were trying to catch or did catch. I look back at the mighty oak tree and notice pine trees in the background and squirrels scampering around scavenging for nuts. I take another sip of coffee, and I watch the Lark Sparrows and Brown-headed Nuthatch birds flying from branch to branch and chirping, while up in the pine trees, I see some Red-cockaded Woodpeckers pecking away looking for a meal. What a beautiful morning. It's nice to get out in nature and enjoy the beauty of nature!

If it's the right time of the year, I will see Robins scouring the grassy areas looking for juicy earthworms to gobble up. Speaking of juicy earthworms, they are my favorite bait when fishing for Big Mouthed Bass. When they strike, it's a good fight to bring them in. I like to cast out among the lily pads, gently bring the worm up one side of the lily pad across the top, and then drop it back down into the water on the other side. I give it a small jerk or two to tease the bass and then pull it onto the next lily pad. If I do it right, I will entice a nice bass to explode upward, breaking the surface of the water and swallowing the worm whole. Sometimes I will take some fat crickets to try and hook a bluegill (also known as Brim) or bass. If I am down at the Cape Fear River fishing for catfish, I will use chicken livers or crawfish, depending on what is available. The catfish go crazy for these, and I never go home empty-handed.

Some days, I walk along the river's steep banks, sightseeing, and enjoy being outdoors. One must be careful to look out for poison ivy among all the foliage among the trails. As I look down before stepping over an old dead fallen tree, I see mushrooms growing on the decaying part of the tree and a colony of termites hard at work eating the dead tree. It is amazing that these insects eat wood, which helps clean up the forest, just like the dung beetle, which eats poop left by animals and lays its eggs in poop, so its offspring has a meal when they hatch. How cool is it that nature has its own cleanup crew? I must be careful with all this sightseeing because I have come across a few Yellow Jacket nests in the past. The nest has small holes in the ground where you will see the yellow jackets entering and leaving. If you see a nest, do not disturb them because they are very aggressive, and their sting is very painful. Unfortunately, I accidentally disturbed a nest and felt the sting. Ouch! As I walk along and look down to the water's edge, I can see turtles warming themselves in the sunshine on a half-submerged log in the water, which reminds me to be on the lookout for Copperhead snakes lurking about. Luckily, I am just west of alligator territory, so I am not really concerned about seeing one of those today. I do from time to time see the fileted carcasses of fish from a previous fisherman or a dead deer full of flies, maggots, and insects feasting with Vultures circling overhead waiting to come down and take part in the feast. Again, another cleanup crew of nature helping to keep everything tidy and neat. Unless you are in the woods in winter or fall, I would suggest some form of insect repellent because, in North Carolina, the gnats and mosquitoes can be very annoying, and the repellent should protect you from ticks and other insects.

The trails along the Cape Fear river bank and some of the State Parks and camping areas here in North Carolina offer the chance to see some great wildlife. I have been lucky enough to see foxes, coyotes, deer, rabbits, and even black bears. I have also had the opportunity to see a Bald Eagle and its nest out by the coast. It is amazing to see these creatures out in nature.

Growing up on the east coast, my father would take my brother and me into some of the marshlands of the intercoastal waterways, where we would go "clamming." That was a lot of fun. We would get a 5-gallon bucket and walk in the muddy tidal areas searching for clams. When we found a clam, we would stick our hands down in the mud and fight with the clam to bring it to the surface. We would sometimes turn it into a little competition to see who could get the most to fill up our bucket. Dad also took us to gather oysters. At the end of the day, dad would show us how to shuck the oysters. Then we would eat the raw oyster on a cracker with a little hot sauce. Wow, that was good! We always steamed the clams and ate them with either butter sauce or a little seasoning. Those are some fond memories, and you cannot beat the taste of fresh oysters and steamed clams. He also took us fishing for Croaker at the ocean on a pier or at the inlet, and the Seagulls and Pelicans would gather near the spot where we would clean the fish to fight for the scraps that we would throw to them. Sometimes we would be in the right area to see schools of

Mullet jumping out of the water, trying to outrun the Snook that was chasing them. We would walk the shoreline on most days at the beach and watch the Sandpipers running along the swash zone feeding on Sand Crabs.

Depending on the time of year, we would see thousands of Portuguese man o' war washed up on the beach. The Portuguese man o' war is such amazing marine life. They look like a blue-purple bubble floating along with long bumpy strings hanging off the bottom of the bubble, and yes, the sting from the tentacles is very painful. Believe it or not, the Loggerhead Sea turtles consider munching on this marine life a delicacy. We would go out on my cousin's boat on a few occasions, where we saw the dorsal fin of a shark or two. I was fascinated as a kid when I learned that sharks don't have bones. Sharks are all cartilage, except for the scary part, the teeth. Yikes! How crazy is that! No bones except for the teeth and a skeleton made entirely of cartilage. When we were out on the boat, dolphins would race the boat and jump out of the water, crisscrossing in front of the bow. That was always an awesome sight to see, as was the time I was snorkeling and saw Stingrays. Stingrays are so graceful and just glide through the water.

Nature is awesome, all the variety of life and how it all works together. It is amazing how everything out there seems to have a true purpose in the hierarchy of nature when you look closely. Even mosquitos. Turns out there are more than 3,500 species of mosquitoes. WOW! (Google search 12 July 2019) Not only are they food for many creatures, but they also help pollinate. Very interesting.

Think about this, the mosquito has over 3,500 species, and only about 150 species of mosquitoes bite humans. Of those, it's only the female mosquitoes that bite. That is amazing.

That got me thinking about Evolution. Is evolution real or made up? Did the human species and all the life that I just described, all the life that we see in the world today, did it all evolve from the first life form, a single cell? Is evolution really the answer? Where did we come from? Hmm, good question.

CHAPTER 1

The Theory of Evolution

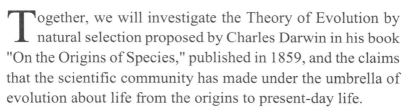

Together, we will investigate the Theory of Evolution by natural selection proposed by Charles Darwin in his book "On the Origins of Species," published in 1859, and the claims that the scientific community has made under the umbrella of evolution about life from the origins to present-day life.

To do this, we must first define what a "theory" is and understand how the term "theory" is defined and used in the scientific community. Then we will all be speaking the same language, so to speak, as we progress with our investigation. According to the website [1]

a theory is defined as; *"In science, a theory is an attempt to explain a particular aspect of the universe. Theories can't be proven, but they can be disproven. If observations and tests support a theory, it becomes stronger, and usually, more scientists will accept it. If the evidence contradicts the theory, scientists must either discard the theory or revise it in light of the new evidence."* The Merriam-Webster dictionary defines theory as *"3.b An unproven assumption."* From the Oxford Languages, a theory is defined as *"a supposition or a system of ideas intended to explain something, especially one based on general principles independent of the thing to be explained." "Darwin's theory of evolution"*

---

[1] https://science.howstuffworks.com/dictionary/astronomy-terms/big-bang-theory.html

Great, and now you and I understand the scientists and will move forward with the understanding that theories are observations, assumptions, or attempts to explain something. Theories are not facts, but the theory can be made stronger or weaker with continued scientific testing and observations.

Now, what are Darwin and the other scientists claiming in the Theory of Evolution and the formation of life?

Darwin's book introduced the scientific theory that populations evolve over the course of generations through a process of natural selection. The book presented a body of evidence that the diversity of life arose by common descent through a branching pattern of evolution. [2]

"Darwinism is a theory of biological evolution developed by the English naturalist Charles Darwin (1809–1882) and others, stating that all species of organisms arise and develop through the natural selection of small, inherited variations that increase the individual's ability to compete, survive, and reproduce. Also called Darwinian theory, it originally included the broad concepts of transmutation of species or of evolution which gained general scientific acceptance after Darwin published "On the Origin of Species" in 1859, including concepts that predated Darwin's theories." [3]

Wikipedia's definition of natural selection "is the differential survival and reproduction of individuals due to differences in phenotype." [4] states natural selection as "the process through which species adapt to their environments. . . traits that give them some advantage are more likely to survive and reproduce." I will expound more on natural selection later in the book.

---

[2] https://en.wikipedia.org/wiki/On_the_Origin_of_Species
[3] https://en.wikipedia.org/wiki/Darwinism
[4] www.nationalgeographic.org

To be fair, Charles Darwin and a lot of the scientific community are not suggesting that evolution/ natural selection originated life. But from that very moment that life began, evolution/ natural selection has been at work, evolving life into what it is today. It is suggested that life came to be on Earth by the process known as Abiogenesis.

Wikipedia explains "abiogenesis" as "the origin of life is the natural process by which life has arisen from non-living matter, such as simple organic compounds. *While the details of this process are still unknown*, the prevailing scientific hypothesis is that the transition from non-living to living entities was not a single event but an evolutionary process of increasing complexity that involved molecular self-replication, self-assembly, autocatalysis of cell membranes."

If the details of the process of abiogenesis are unknown, then how has it been accepted as absolute fact? How is this unknown process taught in public schools as the only factual explanation of how life originated on planet Earth? If I proclaimed as *fact* that aliens populated the earth and then called on lawmakers to outlaw public schools from teaching any other interpretation of the origin of life but admitted that I did not know the details of how the aliens populated the Earth, I would be ridiculed, destroyed and my credibility would be worthless. I hope you see the irony here. This would be my first common-sense question, why is evolution taught as fact when the details of the process are still unknown?

Now we have the big picture of what evolutionists are claiming (claiming an unknown process as fact) through the processes of abiogenesis, evolution, and natural selection. Scientists claim the earliest undisputed evidence of life on Earth dates to about 3.6 billion years ago and that the Earth is about 4.6 billion years old. If scientists are stating that the

Earth is 4.6 billion years old, do we know how the planet formed, what did early Earth look like, and what were the conditions like on the planet at that time? Were these conditions favorable to bring forth life? To better understand how our planet was formed, we need to go back and look at how our Solar System was formed, and to do that, we need to go all the way back to the beginning of time. Time Zero. That very first instant, the universe came into being. Now, you are probably asking, why do we have to go back to that very first moment in time? Is it even possible to get to that point, Time Zero? Do scientists have any theories for how the universe started or what started the universe? Or how old is the universe? Yes, they do.

According to scientists, theoretical physicists, and astronomers, the universe was created approximately 13.8 billion years ago. Life is said to have originated 3.6 billion years ago. If we did not go back to Time Zero, we would be saying those first 10.2 billion years, or basically, 75 percent of the history of time does not matter. How foolish would that be? We must investigate from the very beginning to see what hidden treasures, secrets, or mysteries might be waiting to be discovered. Could they hold the answer as to whether or not evolution is the answer for how we came to be?

CHAPTER 2

Time Zero

There are multiple ideas on how the universe was created. The following is not an all-encompassing list of all the theories but simply some of the more notable theories that are proposed and debated in the scientific community. They are as follows in no particular order: the Big Bounce Theory, White Hole Theory, Inflation Theory, Chaotic Inflation Theory, and the Big Bang Theory. Don't worry; you do not need to be an expert on these theories, but it is important to understand each theory's explanation for how the universe began. With that, here is a general description of each theory.

The Big Bounce Theory builds on Hubble's Law (Interpreted in the simplest fashion, the Hubble law implies that 13.8 billion years ago, all the matter in the universe was closely packed together in an incredibly dense state and that everything then exploded in a "big bang,"). A simplified explanation of The Big Bounce Theory is that it's built on that of the Big Bang Theory by suggesting that the universe contracts to that incredibly dense state and then "bounces" expands outward until it reaches a point and begins to contract again. Some who support the Big Bounce say there have been multiple bounces, while some believe there has only been one bounce.

The White Hole Theory suggests that a White Hole birthed the universe. A White Hole is the hypothesized opposite end of a Black Hole. Everything that a Black Hole has swallowed up and condensed down to a singularity would explode out with great force on the opposite side, the White Hole. In the White Hole Theory, our universe exploded out of a White Hole. There is much debate on whether White Holes actually exist.

"Inflation Theory comes into play in the earliest time measurements in the Big Bang Theory. The inflationary universe is identical to the Big Bang universe for all time after the first 10^{-30} seconds. Prior to that, the model suggests that there was a brief period of extraordinarily rapid expansion or inflation, during which the scale of the universe increased by a factor of about 10^{50} times more than predicted by standard Big Bang models. The Inflation Theory helps explain the remarkable uniformity of the universe and why the universe's density is equal to the critical density."[5]

Chaotic Inflation Theory proposed by physicist Andrei Linde is a variety of Inflation Theories. Chaotic Inflation says our universe is one of many universes that grew as part of a multiverse. This multiverse happened due to a vacuum that had not yet decayed to its ground state. In this theory, the peaks in the evolution of a scalar field (determining the energy of the vacuum) correspond to regions of rapid inflation which dominate, creating "bubble universes."

The Big Bang Theory is one of the most accepted accounts for how the universe began. Although Georges Lemaitre, the Father of the Big Bang Theory does not try and explain what initiated the creation of the universe. The Big Bang Theory does try to explain how the universe developed 13.8 billion years ago from a singularity into what we now observe. A singularity is a point of zero volume and infinite density. All

---

[5] (https://courses.lumenlearning.com/astronomy/chapter/the-inflationary-universe/)

matter, energy, and space expanded outward in all directions very rapidly from that singularity and continue to do so today.

All the theories tend to have certain similarities. For example, all the theories state that the universe is still expanding to this day and that at some point in our universe's development, depending on the theory, it either started as a singularity or contracted back to something very close to a singularity. The most important takeaway here is that ALL the scientists know that the universe isn't just here with us living in it. The universe is the result of an action. Something brought the universe into existence, and we are part of that creation.

Now since the Big Bang theory is one of the most accepted accounts on how the universe began, this is the theory we will be referring to when discussing and asking common-sense questions about the beginnings of the universe some 13.8 billion years ago.

All the matter that makes up everything in the entire universe was released during the Big Bang. What is matter, you ask? Here is Wikipedia's definition of matter.

"In classical physics and general chemistry, matter is any substance that has mass and takes up space by having volume. In everyday, as well as scientific usage, 'matter' generally includes atoms and anything made up of them. Matter exists in various states (also known as phases) such as solid, liquid, and gas"[6]

All of this matter is made up of atoms, and atoms are composed of different combinations of three subatomic particles. These three subatomic particles are the Proton, Neutron, and Electron.

---

[6] https://en.wikipedia.org/wiki/Matter

The first mystery of the creation of the universe are these three subatomic particles, for without them, nothing would be possible. All protons are uniform in size and shape with a positive electrical charge. Neutrons are uniform in size and shape with a neutral charge, and electrons are uniform in size and shape but with a negative electrical charge. These characteristics allow them to bond together to form atoms. These atoms form the different elements on the periodic table, like Helium and Hydrogen, which were formed in abundance at the moment of the Big Bang.

That is just jaw-dropping amazing! In that first instant of utter chaos, these three different subatomic particles are all made to absolute exact standards with zero defects. Scientists have never found one of these subatomic particles that are not identical to its respective kind. Can you even begin to imagine the insane odds of every single subatomic particle being created absolutely perfect? Wow!

On top of that, what caused these three perfect subatomic particles to carry an electrical charge? Why doesn't this electrical charge ever diminish? These particles keep their respective charge forever. That is amazing! How did the electrical charge even come to be at that instant?

Right from the singularity in that first infinitely small spec of time, these three subatomic particles with their perfect relationship to bond was created. If there is no electrical charge, or all positive, all negative, or all neutral, there is no bonding of these particles. No bonding means no atoms. No atoms mean no elements and no existence. Just infinite space filled with subatomic particles that cannot bond. It is almost like the formation of the universe was planned.

CHAPTER 3

The Universe

———•○◇○•———

Time Zero, the singularity. The Big Bang happens, and time begins. At the moment of the Big Bang, ALL matter, space, and energy exploded outward uniformly in all directions. What did that expansion look like? In very simple terms, as the universe began expanding, there were massive clouds of dust and gasses that were many light-years across. According to scientists, these clouds are what will eventually form the sun and planets of our solar system. Some of these clouds can still be seen today, and these clouds may still produce many other stars, planets, and galaxies within the universe. Wikipedia states that dwarf galaxies can have "just a few hundred million stars to giant galaxies with one hundred trillion stars." Astronomers are confident in stating that there are between 100 billion and 200 billion galaxies in our universe. This is the universe we are a part of. That is simply awe-inspiring to imagine all that matter and energy was condensed into a singularity, a point with zero volume and infinite density. Wow, just wow! This universe is not by accident; our universe was brought into existence.

The Big Bang happened, and now there are these massive clouds of dust and gasses expanding outward that will form stars and planets, but how? Did one of these clouds form our solar system? Good question.

Here is how the scientific community explains the process. The cloud will form a star first, and then one or more planets will form from the left-over matter of the cloud. The process starts when energy waves pass through these clouds, creating turbulence deep within the clouds that form knots. *Do these energy waves start the rotation of the cloud? If not, what causes the rotation of these clouds? Without rotation, there are no planets.* These knots become high-density regions that contain enough mass that the gasses and dust begin to collapse due to the gravitational attraction. As this process happens, the gravitational force causes the material at the center to heat up to very high temperatures creating a protostar. Eventually, the temperatures at the center become hot enough to ignite fusion. When fusion has been achieved, a star is born.

To give you a mental picture of what this process looks like, imagine a giant puffy cloud, and then deep in the center of that cloud, the knot is formed. The knot starts condensing the puffy cloud at the center, which increases its gravitational forces. This process causes the cloud that is rotating around the knot in the center to flatten like a pancake. As the knot continues to pull dust and gasses into the center, it becomes denser and increases in size. Soon it will look like a round ball with flat rings around the ball. Imagine Saturn and its rings, but in this scenario, the planet would be the star, and the rings would stretch out much further. The dust and gasses in these rings will start to collide as they rotate around the star in the center until all the matter in its path has become part of the planet. Now you have a star with one or more rotating planets. This is also the explanation given for the formation of our solar system. This process could be correct, but there are still some unknowns, at least about how stars are formed, as shared by the following quote from the ALMA Observatory.

"Stars shine for billions of years, but their formation – which takes only a few million years – remains, literally, a mystery: optical telescopes cannot observe inside the dusty clusters of gas where stars are born. On the other hand, infrared telescopes that can reveal infant stars before they completely emerge from their dusty cradles, are not able to observe the development process involved in the pre-ignition of stars." [7]

Google "how are stars formed." My search returned 2,000,000,000 results in less than a second. All the articles I have read, whether from this Google search or other research, all plainly state the process by which a star is formed as undisputed fact, except for this quote from the ALMA. Scientists have a strong hypothesis on the formation of stars, according to an observatory that is adhering to scientific principles and reporting it honestly. However, there is still some crucial information missing. This is how scientists are supposed to present their findings. Scientists observe, form hypotheses, and then try and prove them. If unknowns are discovered, that is ok and should be expected as part of the scientific process. But these unknowns must be plainly stated. Scientists MUST stay true to science! When scientists get involved with promoting agendas or politics, they lose the public's trust and their credibility as scientists. They are now nothing more than paid propaganda and should be treated as such. Having personal opinions about a theory is fine, but a scientist cannot let that personal opinion corrupt their research.

Armed with this new knowledge that there might still be some mysteries about exactly how our solar system was

---

[7] https://www.almaobservatory.org/en/about-alma/how-alma-works/capabilities/star-and-planet-formation/

formed, I have a few more questions that I have not been able to find the answers to yet. And I am not sure if the present capabilities of science can honestly answer them. Questions like, in our solar system, if planets are formed from the left-over rings of dust and gasses rotating around the star, how do these dust and gas particles start colliding and clumping together if they rotate at the same speed in the same direction? For argument's sake, these particles start clumping together and growing in size and mass. How does the growing planet with a growing mass stay in the same orbit as the small particles with very little mass? If the sun's gravity was holding the small particles with very little mass in a particular orbit, how is this growing planet with a growing mass staying in the same orbit?

Maybe I need to do more research on gravity and its effects. Every non-fiction source explains the force of gravity the same way so let's take an article from Sciencefocus.com that explains the force of gravity. "What is gravity? It's a property of matter, of stuff. In a nutshell: all matter is attracted to all other matter. The more matter there is, and the closer objects are to each other, the bigger that attractive force. And unlike electricity and magnetism, which can either repel or attract, gravity always pulls things together." This article pulls from Newton's theory of gravity which explains that "the attractive force of gravity goes up as either of the objects' masses increases or as they get closer together."[8]

And the definition of gravity from Wikipedia says this.

"Gravity (from Latin gravitas 'weight'[1]), or gravitation, is a natural phenomenon by which all things with mass or energy, including planets, stars, galaxies, and even light,[2] are attracted to (or gravitate toward) one another. On Earth,

---

[8] https://www.sciencefocus.com/space/gravity/

gravity gives weight to physical objects, and the Moon's gravity causes the tides of the oceans." [9]

Gravity pulls objects together. That is how the star is formed from the dust cloud, and the gravitational pull of the star is strong enough to keep the rings of particles rotating around the star but not strong enough to pull them closer. Ok, I can understand that, but as the particles start to clump together to grow this new planet and as the mass of this new planet is increasing, why is it not getting pulled closer to the star? Look at what NASA says about Neptune and Uranus with that question in mind.

Neptune: Neptune took shape when the rest of the solar system formed about 4.5 billion years ago when gravity pulled swirling gas and dust in to become this ice giant. Like its neighbor Uranus, Neptune likely *formed closer to the Sun and moved to the outer solar system* about 4 billion years ago. [10]

Uranus: Uranus took shape when the rest of the solar system formed about 4.5 billion years ago – when gravity pulled swirling gas and dust in to become this ice giant. Like its neighbor Neptune, Uranus likely *formed closer to the Sun and moved to the outer solar system* about 4 billion years ago, where it is the seventh planet from the Sun. [11]

Both planets formed closer to the sun, and then, as each planet grew in size or once they had matured in size, moved away from the sun? Does that make sense? In our solar system, Uranus and Neptune, the third and fourth largest planets, respectively, were able to overcome the suns and their own gravitational forces and move away from each other? How is this possible when according to Newton's theory of gravity, gravity *Always* pulls objects together, and the

---

[9] https://en.wikipedia.org/wiki/Gravity
[10] https://solarsystem.nasa.gov/planets/neptune/in-depth/
[11] https://solarsystem.nasa.gov/planets/uranus/in-depth/

attractive force of gravity goes *Up* as *Either* of the objects' *Masses Increase* or as they get *Closer* together!

Like a check on learning in a textbook, let's have a check on what we are being told. Small particles in the dust ring are held in place by the sun's gravity. Check that makes sense. The growing planet is also held in the same orbit by the sun's gravity with its growing mass. Ah, wait, that seems a bit contradictory to the definition of gravity. The 3rd and 4th largest planets in our solar system formed and were held in close orbit to the sun by its gravitational force but then moved to a new orbit in the outer solar system. Just not possible according to Newton's theory of gravity.

Gravity plays another role in the formation as well as the shaping of planets, according to scientists, as stated in this article from Scientific America.

"Planets are round because their gravitational field acts as though it originates from the center of the body and pulls everything toward it. The only way to get all the mass as close to the planet's center of gravity as possible is to form a sphere."[12]

The particles start clumping together, and the growing planet has a growing mass that increases its gravitational force. This force pulls in all directions into the center, the core, of the growing planet. This is what causes the planet to take on the shape of a sphere.

But what is still not answered is how the growing planet stays in the same orbit as the small particles that it is gathering up. And how did Uranus and Neptune form closer to the sun, yet against the sun and their own gravitational force move to the outer solar system and through the pathways of other forming planets? In this article from Discover magazine, Renu

---

[12] https://www.scientificamerican.com/article/why-are-planets-round/

Malhotra, a staff scientist at Houston's Lunar and Planetary Institute, suggest (theorizes)

(Neptune— a planet with a mass of more than 17 Earths— influenced the orbits of all its diminutive neighbors. After the planets formed, their gravitational pull interfered with the orbits of left-over building blocks called planetesimals. "What evolved was a sort of planetary game of handball, involving Neptune, Uranus, Saturn, and Jupiter," says Malhotra. By the rules of this interaction, Neptune literally tugged planetesimals from their orbits and "handed them down" to the giant gas planets closer to the sun. As the planetesimals moved into smaller orbits, they lost orbital energy and angular momentum— and that energy was instead absorbed by the planets. "The greater the orbital energy, the bigger the orbit," explains Malhotra. Neptune, Uranus, and Saturn began to move outward— with Neptune taking a 30 percent leap that moved it to its present location of roughly 2.8 billion miles from the sun. Jupiter, which lay on the receiving end of the planetesimals, either absorbed or ejected them. As a result, Jupiter lost orbital energy, and its circuit around the sun shrank by about 2 percent.) [13]

Once Neptune altered that planetesimal's orbit, the corresponding energy was also altered. How does Neptune absorb the energy that the planetesimals lost? If Neptune's gravitational pull did actually overcome the angular momentum of the planetesimals, I do not see how that energy can be transferred to Neptune. For sports fans, if Tom Brady threw a pass to his receiver but missed him, does the receiver gain energy from the ball slowing down as it goes by? If the receiver caught the ball, would the receiver run faster by absorbing the energy that is propelling the ball? Also, let's not

---

[13] https://www.discovermagazine.com/the-sciences/neptune-rising

forget Newton's theory, "the attractive force of gravity goes up as either of the objects' masses increases or as they get closer together." If Neptune was getting close enough that it could knock these planetesimals from their orbits, why wouldn't the two objects be drawn together by their increasing attractive force? I understand that angular momentum could be stronger than the gravitational pull of these two objects, resulting in the planetesimal being slung out of orbit in some direction but knowing which direction it would be slung and the transfer of its energy to the other planets does not seem logical. And if it did actually happen, how many of these smaller planetesimals would it take to give Neptune, Uranus, and Saturn enough "orbital energy" to break free from their original orbits and move further outward, knocking Neptune 30 percent further outward?

Now the website here suggests why planets orbit a star. The explanation creates a few questions that I put in bold type. [14]

"The reason the planets are traveling at just that speed which allows them to orbit the Sun (and not spiral into it or whirl away into space) is not a coincidence or evidence of divine intervention but goes back to when the Solar System was just a spinning cloud of gas and dust. Everything that was spinning slowly was incorporated into the Sun itself under the force of gravity; everything that was spinning too fast escaped into outer space; everything else remained in orbit around the Sun and gradually coalesced into the planets, retaining its speed of spin and therefore its orbit (encountering little resistance in the near-vacuum of space)."

My question here is, as the mass of the coalescing planets increased, wouldn't its gravitational force also increase? Yes, it would; remember Newton's theory. That means if the

---

[14] https://www.physicsoftheuniverse.com/what-do-the-planets-orbit-the-sun.html

gravitational force increases, the speed required to keep the planet in orbit has to increase; otherwise, according to this very article, it would be orbiting too slowly and be pulled into the star. And what about Neptune, Saturn, and Uranus? These planets were also in an established orbit, which means they were "traveling at just that speed which allows them to orbit the Sun (and not spiral into it or whirl away into space)." These three planets had to increase the speed of their individual orbits to gain enough momentum to break free of their individual original orbits and move further away from the sun. The question is, once the planets were able to break free of the original orbit and start moving away from the sun, how did that momentum not cause these planets to "whirl away into space?" Neptune moved 30 percent further away from the sun. Once these planets started moving outward, what was able to overcome the planets' new momentum and hold them in their new respective orbits?

Now with regards to Jupiter:

"If the mass of all of the other planets in the solar system were combined into one "super planet," Jupiter would still be two and a half times as large."[15]

Jupiter has two and a half the mass of all the other planets combined in our solar system. This would mean that Jupiter has the greatest gravitational force of all the planets in our solar system. If absorbing or rejecting these planetesimals caused Jupiter's orbit to slow enough that it dropped out of its orbit and began moving closer to the sun, even by only 2 percent, the gravitational force between the sun and Jupiter would increase as Jupiter moved closer to the sun. The "physicsoftheuniverse.com" article says that "the planets are traveling at just that speed which allows them to orbit the Sun

---

[15] https://www.space.com/18392-how-big-is-jupiter.html

(and not spiral into it or whirl away into space)." According to this article, if you alter that speed, the planet would either spiral into the sun or whirl away into space. Jupiter's momentum slowed enough for its orbit to shrink by 2 percent. That is huge! So why did it not spiral into the sun? And why did Neptune, Uranus, and Saturn whirl away into space? The common-sense question is, how do scientists explain this?

"Because the Sun and planets all formed from the same spinning nebular cloud, this is also why they all rotate in the same direction. As the nebula continued to contract under the influence of gravity, it rotated faster and faster due to the conservation of angular momentum. Centrifugal effects caused the spinning cloud to flatten into a flattish disk with a dense bulge at its center (which would coalesce into the Sun). This is why the planets orbit the Sun in a more or less flat plane, known as the ecliptic." [16]

If gravity is trying to pull everything into the center from all directions equally, causing the bulge at the center of the cloud and the spherical shape of stars and planets, how is gravity creating rotation? Centrifugal force can only begin after the rotation has started. What started the initial rotation of the cloud? Why is the speed of rotation for the planets and the sun constant and infinite? Shouldn't it either be speeding up or slowing down by the laws of physics?

The article also contributes to the increase in rotational speed due to the conservation of angular momentum. If you are not familiar with this theory, do not worry; neither was I. Here is a simple example to help you visualize the conservation of angular momentum. Think of an ice skater spinning around on one skate with their arms stretched out. As they pull their arms in, the rotation speeds up. Tada, that's it

---

[16] https://www.physicsoftheuniverse.com/what-do-the-planets-orbit-the-sun.html

in a nutshell. Cool, right. Although, that brings up another question with regards to the solar system and planets. Stick with me for a moment. Hypothetically, if the skater's hands were not attached to the arms, would the speed of rotation for the hands speed up as the arms were pulled into the skater's body?

This same example should apply to the spinning nebular cloud. The conservation of angular momentum is causing the center of the cloud to spin faster as it draws more matter in towards the center. Still, the rest of the cloud should continue to spin at the original speed because there is no physical connection. The skater's hands were physically connected to the skater, so yes, as the arms and hands were drawn inward, the speed of rotation for the skater and the hands increased. Next, this article states that centrifugal force causes the cloud to flatten with the bulge at the center. Can centrifugal force act on matter that is not connected to each other? When you go to the fair and ride one of the spinning rides, like the "Gravitron," you are standing against the outer wall, so centrifugal force pins your body to the wall. But if you were floating in between the center and outer wall of the ride without actually touching the ride, would centrifugal force affect you? But continuing with the argument that centrifugal force does form the flattened disk around the star, the particles in the disk should start small at the center of the rotating disk and increase in size and mass as you move to the outer edge of the disk. But in pictures, the rings at different distances from the sun have objects of different sizes in them. How are smaller objects held in the outer rings without being flung out into space and bigger objects in the inner rings not pulled into the star?

CHAPTER 4

Our Solar System

The universe is 13.8 billion years old, and it took 9.2 billion years to form our solar system. Let's look at our solar system and the eight planets that orbit around our sun. In an order from closest to farthest, they are Mercury, Venus, Earth, Mars, Jupiter, Saturn, Uranus, and Neptune. I did not include Pluto because it is no longer recognized as a true planet. Jupiter is the biggest planet by far, and Mercury is the smallest in our solar system. Earth is the fifth largest planet. Jupiter has a mass more than two and a half times that of all the other planets in our solar system. According to NASA, more than 1300 earths could fit into Jupiter. Now that is a BIG planet! WOW! But when you compare Jupiter to the sun, Jupiter comes up very short. "You could fill the Sun with 1,000 Jupiter-sized planets."[17] Just a bit of fun trivia for you.

Now when we compare all the planets with regards to the time it takes to make one rotation on the planet's axis and one orbit around the sun, Earth seems to have the best, yet I say, the perfect combination to support life. Check this out.

Earth makes one complete rotation around its axis in 24 hours and takes 365 days to rotate our sun with an axial tilt of 23 degrees. This axial tilt is responsible for Earth's seasons. When compared to the other planets in our solar system, Earth

---

[17] https://nineplanets.org/questions/how-big-is-jupiter/

seems to have the best, perhaps the perfect rotation, orbit, and axial tilt to support life. By comparison, Venus takes 243 earth days to complete one rotation on its axis and 225 days to orbit the sun with an axial tilt of only 2.64 degrees. (One day on Venus is longer than one year on Venus.) Neptune takes only 16 hours to complete one rotation on its axis but 165 earth years to orbit the sun. Mars, like the Earth, completes one rotation on its axis in 24 hours but takes 687 days to orbit the sun. By random chance, Earth just happens to be at the correct distance from the sun to sustain life. By random chance, Earth just happens to have a rotation, orbit, and axial tilt that provides the best possibility of supporting life. By random chance, Earth has only one moon. Fun fact, Earth's moon is the fifth largest moon in our solar system, and Earth is the fifth largest planet in our solar system. Also, the "Moon makes Earth a more livable planet by moderating our home planet's wobble on its axis, leading to a relatively stable climate. It also causes tides, creating a rhythm that has guided humans for thousands of years."[18]

It is truly incredible that just to create a platform that could and will eventually support life, everything has been perfectly made and placed in the right place without a single flaw. And remember that none of anything is possible without protons, neutrons, and electrons assembled in different combinations to create every element, which in turn is the foundation of building everything else in the ENTIRE universe. How perfect. The question of whether this is the result of random chance (and eventually evolution through natural selection) or a Creator must be entering your mind by now. And it only gets better as we progress.

---

[18] https://solarsystem.nasa.gov/moons/earths-moon/in-depth/

Earth

The Theory of Evolution started 9.2 billion years after the moment when the universe came into existence. Our planet Earth, the third rock from the sun, formed 4.6 billion years ago, and the first life formed 3.6 billion years ago. What were the conditions like in that first billion years on Earth? From some of the earlier sources, Earth was in a stage referred to as the "primordial soup."

Dictionary.com defines the primordial soup as "the seas and atmosphere as they existed on earth before the existence of life, consisting primarily of an oxygen-free gaseous mixture containing chiefly water, hydrogen, methane, ammonia, and carbon dioxide."[19]

According to the Smithsonian Environmental Research Center (SERC). This ancient atmosphere was very different from today's; it had hydrogen sulfide, methane, and 10 to 200 times as much carbon dioxide as the modern atmosphere does, according to SERC. "We believe the Earth started out with an atmosphere a bit like [that of] Venus, with nitrogen, carbon dioxide, maybe methane," said Jeremy Frey, a professor of physical chemistry at the University of Southampton in the United Kingdom. "Life then began somehow, almost certainly in the bottom of an ocean somewhere."[20]

Here is a fair, common-sense question. If the possibility of random chance creating life on Earth is so easily accepted and Earth started out with an atmosphere like Venus, our neighbor, then why didn't random chance create life on Venus as well?

Scientists have told us what the atmosphere was like in the primordial soup of early Earth, but do scientists know some of

---

[19] https://www.dictionary.com/browse/primordial-soup.
[20] https://www.livescience.com/64825-why-earth-has-an-atmosphere.html

the other conditions of the primordial soup? Was early Earth hot or cold? Ok, it was "As warm as it is today." [21]

Or maybe hotter, as this article says: "Even after collisions stopped, and the planet had tens of millions of years to cool, surface temperatures were likely more than 400° Fahrenheit." [22] Oh, hold on. Scientists also claim early Earth was cold. "We have found evidence that the climate 3.5 billion years ago was a cold environment," "This may indicate that Earth, 3.5 billion years ago, experienced an extensive, perhaps global, ice age," says Furnes." Harald Furnes, Professor Emeritus, the Department of Earth Science, University of Bergen. [23]

This article also claims that the early earth was very cold. "When Earth's first organisms were formed, it may have been in an ice-cold ocean. New research, published in Science Advances, indicates that both land and ocean were much colder than previously believed." [24]

Two articles claim the early Earth was Hot, and two articles claim the early Earth was Cold. Scientists cannot even agree on just the physical conditions of the primordial soup of early Earth, but that same scientific community demands that we accept Evolution, the origins of life, and natural selection as proven facts. And for those of us who question the so-called "facts," it is said that we just cannot accept proven science. In my opinion, science points to a Creator, not to random chance. With that said, let's get back to early Earth.s

In this primordial soup, 3.6 billion years ago is where Scientists and evolutionists believe life happened by accident. Charles Darwin and others claim that natural selection came

---

[21] https://www.nationalgeographic.com/science/article/130715-ancient-earth-science-sun-paradox-evolution.
[22] https://www.climate.gov/news-features/climate-qa/whats-hottest-earths-ever-been
[23] https://partner.sciencenorway.no/climate-forskningno-norway/early-earth-may-have-been-freezing-cold/1431293#
[24] https://www.sciencedaily.com/releases/2016/03/160317144620.htm

into play as soon as life accidentally came to be. Random chance started life, but then natural selection took over? How does this pass the common-sense test? From inorganic or inanimate substances that by random chance combine to form the building blocks of life with the inherent desire to start evolving. That seems like a pretty big stretch of the imagination. So, what was the probability of life forming by random chance in the primordial soup? The mathematical probability of life, and not even true life, just a single protein chain-forming by random chance, is truly mindboggling, regardless of who's answer you believe.

Chapter 5

Probability of Life (Amino acids & Proteins)

Here are excerpts from two articles of opposing theologies that both speak on the mathematical likelihood of life forming by accident/ chance on the newly formed planet earth.

The first one is a video supporting Creationists: (Origin: Probability of a Single Protein forming by chance, published April 7, 2017) [25] Which states, "When applied to the origin of life and the random formation of large biomolecules, Probability Theory clarifies the limitations of chance as a creative agent on the primordial earth." This video then sets out to explain its opening statement as follows. In an "ideal environment for chemical evolution, an imaginary world that will provide chance every opportunity to succeed.... stock the ocean to capacity with amino acids, that means all the atoms on earth including its entire supply of carbon, nitrogen, oxygen, hydrogen, sulfur are available to form 10 to the 41st complete sets of the 20 types of amino acids used to build proteins. Then we will alter the laws of nature to protect these building blocks from the destructive rays of ultraviolet light and chemical contamination in the primordial soup.... With all these protections in place, that works out to one correctly

---

[25] https://m.youtube.com/watch?v=W1_KEVaCyaA

sequenced functional protein chain for every 10 to 164th failed attempts (1 right attempt for every 100 million, trillion, trillion, trillion, trillion, trillion, trillion, trillion, trillion, trillion, trillion, trillion, trillion, trillion failed attempts). This calculation is for a smaller than average molecule made from 150 amino acids making up the protein chain." I recommend watching the 9 min 28 sec video. It is very interesting, regardless of what viewpoint you hold. Again, I am only trying to pull out some quick comparison calculations between Creationists and Evolutionists. I will say the last 2 mins 52 secs of the video truly destroyed the idea of life happening by chance.

The second is from the evolutionist viewpoint. [26]

I encourage you to read this article as well. I think it is a great example of the two different viewpoints. To my surprise, as this is just the beginning of my research into evolution, it shows some major assumptions on the part of the evolutionists without any supporting evidence of those assumptions. Here the evolutionist is showcasing the creationist argument before countering with their calculations. "The calculation which supports the creationist argument begins with the probability of a 300-molecule-long protein forming by total random chance. This would be approximately 1 chance in 10 to the 390th power. This number is astoundingly huge. By comparison, the number of all the atoms in the observable universe is 10 to the 80th power. So, if a simple protein has that unlikely chance of forming, what hope does a complete bacterium have?" This article states that the process of life was "all governed by the *non-random* forces of Natural Selection and chemistry." Also stated in this paragraph of this article is that the "first stages. . . were no more than simple self-

---

[26] http://evolutionfaq.com/articles/probability-life

replicating molecules. . ." The evolutionist's viewpoint in this article states that, "the simplest theorized self-replicating peptide is only 32 amino acids long. The probability of it forming randomly, in sequential trials, is approximately 1 in 10 to the 40th power." A peptide is a short chain of amino acids. Yes, you are correct. The evolutionist started the article off by stating this process of *life was non-random* due to the forces of Natural Selection. The next paragraph of the article states the calculation for *the random formation* of this simple peptide.

From the evolutionist view, this article presents the calculations for a subpart of a correctly sequenced functioning protein chain but does not give the odds of forming a complete correctly sequenced protein chain as the Creationist video above does. The evolutionist justification for not calculating those odds is that life happens in many small steps. But what are the odds of these small steps randomly happening in the right order to form the first correctly sequenced protein chain? Was the math too complicated to calculate the odds of those small steps? Or did it not support the theory? This article just assumes it happens after the formation of the first peptide. I find it extremely interesting and perplexing that scientists are just assuming something without providing any scientific evidence to support the assumption that is being offered. The end of this article has a section that reads: *"Please note: this article uses a hypothetical self-replicating peptide as a model molecule for these calculations."*

Wikipedia defines "hypothetical" as: Hypotheticals are possible situations, statements, or questions about *something imaginary rather than something real*. Hypotheticals deal with the concept of "what if?" Just like in the beginning of this book when we defined a theory. A hypothetical model cannot be used to prove a hypothesis.

It is interesting to note the similarities in calculations of the two articles. Evolutionists state the odds of random chance forming a peptide is 10 to the 40th. In the Creationist video, the odds of a 150-chain molecule forming by random chance is 10 to the 164th. The Evolutionist, in their article, claims that Creationists use a 300-chain molecule with the odds of 10 to the 390th of forming by random chance. The evolutionist article even agrees with the Creationists on how unlikely the chance of life forming is if the odds were 10 to the 390th power. The evolutionists are using an old sales trick of throwing up astronomically bad odds and agreeing on how impossible the random formation of life would be at those odds. Oh, but wait, if we cut that number down to a fraction of the 300-chain molecule to only a peptide, the odds of that are much more believable. Now you bought the sales pitch, Hook, Line, and Sinker. You walk away going sure that could happen over a billion years. Ok, I believe. And you never asked what a peptide is. A peptide is a compound of two or more amino acids linked in a chain, but it is not a protein chain. Big Difference.

Now, let's take the evolutionist mathematical probability viewpoint for this debate: a 1 in 10 to the 40th power chance of randomly forming a peptide. 1 in 10 to the 40th power is still extremely bad odds! I mean, that is horrifically bad odds!! And that is just a peptide. Not even a correctly sequenced protein chain.

[27] Here is a fun comparison; The Lightning Safety Council reports your odds of being struck by lightning in a given year was 1 in 1,443,000. Powerball states the odds of winning the

---

[27] http://lightningsafetycouncil.org/Odds.pdf

grand prize is 1 in 292,000,000. Peptide by chance 1 in 10 to the 40th, and finally a Protein by chance 1 in 10 to the 164th.

Struck by lightning: 1 in 1,000,000
Powerball winner: 1 in 292,000,000
Peptide by chance: 1 in 10,000,000,000,000,000,000,000,000,000,000,000,000,000
Protein by chance: 1 in 100,000,000,000,000,000,000,000,000,000,000,000,000,000,000,000,000,000,000,000,000,000,000,000,000,000,000,000,000,000,000,000,000,000,000,000,000,000,000,000,000,000,000,000,000,000,000,000,000,000,000,000,000,000,000,000,0000

Amino acids & Protein molecules

Those are some crazy-looking numbers and, really, bad odds.

In the probability of a protein made up of just 150 amino acids, there are 56 zeros! Now can you even fathom the number of zeros in the probability of a protein by chance if that protein is even just 1,000 amino acids long? Wow, that would be a lot of zeros!

Before these amino acids can start the crazy odds of forming proteins, random chance must first create these amino acids. These amino acids must be plentiful and collocated to begin the protein-making process. The amino acids begin bonding, but the sequence in which they bond is unique for each different amino acid. If the sequence is incorrect, the chain is immediately destroyed, and reassembly begins again. What force is driving these atoms to continuously try different combinations to create different types of amino acids? This process continues until the correct sequence is achieved, resulting in one unique amino acid. Each correct sequence results in one amino acid. Once formed, how long can amino acids survive in the environment of the primordial soup before they are destroyed by the sun? Out of the more than 500 naturally occurring amino acids, only 20 appear in the genetic code. Why? Let's find out what Amino Acids are.

Amino acids are the building blocks of proteins. Every amino acid contains carbon, hydrogen, oxygen, and nitrogen atoms. But first, remember those three little subatomic particles. This is where you start seeing how important protons, neutrons, and electrons are to existence and how everything seems to have been created to satisfy a very specific purpose and role. A hydrogen atom is the simplest atom composed of one proton and one electron, whereas an oxygen atom is composed of 8 protons, 8 neutrons, and 8 electrons. Now you are starting to see how the foundation for the universe, for life, for everything was planned. How everything, even down to subatomic particles, was created with a purpose. All Amino Acids share a general structure composed of four groups of molecules: a central Alpha-carbon with a hydrogen atom, an amine group, a carboxyl group, and a side chain. The general structure for amino acids is the same, but each amino acid sequence is unique to that amino acid. Once the amino acids are sequenced correctly, they fold to make the final shape of the protein. The amino acids have now become a protein.

I didn't ask this earlier because we were simply comparing and discussing the probability of amino acids forming a protein by random chance but now that we know a little more about amino acids, my question is, why did amino acids even form? How or why did carbon, hydrogen, oxygen, and nitrogen decide to link up. They can exist as separate atoms, so why did they bond to create amino acids? Among these are the 20 individual amino acids that are required for life. And why has natural selection determined that we need these 20 amino acids out of the more than 500 amino acids that have been identified? Why not use more or fewer amino acids to support life? And why bond these amino acids in unique arrangements to create proteins? Why are proteins so important in the formation and the continuation of life?

Remember that the odds of 1 correct attempt in 10 to the 164th wrong attempts for random chance to sequence one smaller than average protein chain correctly produces just that, one specific protein chain. That is *only one* correctly formed functioning protein chain. It is also *only one* type of protein.

Here is a little foreshadowing for you in this article. "There are approximately 42 million protein molecules in a simple cell, according to University of Toronto's Professor Grant Brown and co-authors."[28]

That is crazy; 42 million protein molecules in a "simple cell." Wow. What is dictating the different combinations of amino acids to create different proteins? Has science calculated the odds of all the different proteins forming by random chance? How long does that protein chain survive before it is destroyed? Is random chance creating these different proteins just for fun now? Is natural selection taking credit for figuring out what combinations of amino acids are needed to create the proteins that will be needed in the future for the further development of life?

Neither the peptide nor the protein chain constitutes life but is a building block of life. Evolutionists claim that these chemical reactions were happening at a feverish rate all over the planet. That means that the entire planet had the exact same favorable conditions required for the constant attempts

---

[28] http://www.sci-news.com/biology/cell-protein-molecules-05618.html

to produce the needed protein chains. The time needed for all the attempts required to produce these protein chains seems to again discredit the formation of life by random chance. The time required is explained in the YouTube video, cited earlier in this chapter, supporting the Creationist theory. Here are some basic facts about proteins that will be expanded upon later.

Proteins are divided into seven types. Digestive Enzymes, Transport, Structural, Hormones, Defense, Contractile, and Storage [29]

These functions are briefly explained in the Britannica article below.

[30] Reminds us that there are 42 million protein molecules in a cell.

Random chance has now created a protein. In the furthering of life, what is the role and purpose of protein? Why are there 42 million protein molecules in a cell? Britannica explains some of the roles, functions, and capabilities of protein nicely in this article. [31] *Proteins are of primary importance to the continuing functioning of life on Earth. Proteins catalyze the vast majority of chemical reactions that occur in the cell. They provide many of the structural elements of a cell, and they help bind cells together into tissues. Some proteins act as contractile elements to make movement possible. Others are responsible for the transport of vital materials from the outside of the cell ("extracellular") to its inside ("intracellular"). In the form of antibodies, proteins protect animals from disease and, in the form of interferon, mount an intracellular attack against viruses that have eluded destruction by the antibodies and other immune system*

---
[29] https://courses.lumenlearning.com/wm-biology1/chapter/reading-function-of-proteins/
[30] https://www.sciencedaily.com/releases/2018/01/180117131202.htm#:~:text=Summary
[31] https://www.britannica.com/science/amino-acid

defenses. Many hormones are proteins. Last but certainly not least, proteins control the activity of genes ("gene expression").

Some functions of proteins are further defined in this article. [32]

Proteins function as enzymes by carrying out almost all of the thousands of chemical reactions that take place in cells. They also assist with the formation of new molecules by reading the genetic information stored in deoxyribonucleic acid, also known as DNA. Proteins function as messengers, such as some types of hormones, and transmit signals to coordinate biological processes between different cells, tissues, and organs. Proteins also serve as a structural component by providing structure and support for cells. On a larger scale, they also allow the body to move as well as function as proteins that bind and carry atoms and small molecules within cells and throughout the body.

Some different types of protein are collagen. This protein gives structure to your skin, bones as well as your teeth. Integrins are proteins that create links that are flexible between cells. Keratin is a protein found in our hair and nails. Our cells even have a circadian clock protein that keeps time in our cells.

One unique and amazing fact, yes, this is a fact, is that the shape of the protein dictates its role in the cell. Isn't that amazing, the shape that a protein takes dictates its role in the cell. The National Institute of General Medical Sciences describes how a protein's shape is unique to the role and function it performs. [33] *A protein's structure allows it to perform its job. For instance, antibodies are shaped like a Y.*

---

[32] https://medlineplus.gov/genetics/understanding/howgeneswork/protein/
[33] https://www.nigms.nih.gov/education/fact-sheets/Pages/structural-biology.aspx

This helps these immune-system proteins bind to foreign molecules such as bacteria or viruses with one end while recruiting other immune-system proteins with the other. DNA polymerase III is donut-shaped. This helps it form a ring around DNA as it copies its genetic information. And proteins called enzymes have grooves and pockets that help them hold onto other molecules to speed chemical reactions. Misfolded or misshapen proteins can cause diseases. They often stop working properly and can build up in tissues. Alzheimer's disease, Parkinson's disease, and cystic fibrosis are examples of diseases caused by misfolded proteins."

The more we investigate proteins, the more we learn just how important they are in supporting life. We are also learning how unique and specific proteins are. Proteins are not generic molecules needed for life. Each type of protein was designed and created to perform very specific roles. Roles that cannot be performed by any other molecule and roles that, if proteins were not created, life would not be possible. Proteins make it possible for life (life forms) to store and use energy. I would say that's pretty important.

We now have a fair understanding of the life-sustaining importance of proteins, and I am about to show you that proteins are both species-specific and organ-specific within that species.

This article [34] describes proteins in this manner. "Protein is a highly complex substance that is present in all living organisms. Proteins are of great nutritional value and are directly involved in the chemical processes essential for life. . . Proteins are species-specific; that is, the proteins of one species differ from those of another species. They are also

---

[34] https://www.britannica.com/science/protein

organ-specific; for instance, *within a single organism, muscle proteins differ from those of the brain and liver.*"

This article [35] "These findings generated the common belief that the human is extremely close to the chimpanzee at the genetic level. However, if one looks at proteins, which are mainly responsible for phenotypic differences, the picture is quite different, and about *80% of proteins are different between the two species.*"

So how does random chance (supposedly that is what created proteins, according to evolutionists) create protein that is not a part of any species as of yet, that will be species-specific and organ-specific within that species? I am sure the argument is going to be that random chance created generic proteins if that is possible. Then natural selection evolved those proteins over millions of years into unique species-specific and organ-specific proteins. How can evolutionists claim that natural selection knows to evolve that protein into a species-specific protein without a life form? That is honestly like claiming as fact that random selection created the egg and natural selection evolved that egg into a chicken and demanding the world to believe you.

Now that we know the seven types of proteins, their functions, and how specific those roles are. We also know that proteins are actually species-specific, and within that species, proteins are organ-specific. We need to ask how Evolutionists are trying to justify these results by random chance or even by natural selection. One very basic question to ask an Evolutionist would be, is random chance or natural selection still creating proteins? If not, why not? How did natural selection know to create different types of protein and how to fold the amino acid chains creating that unique protein

---

[35] https://pubmed.ncbi.nlm.nih.gov/15716009/

correctly? Did natural selection know that it would evolve diseases and viruses, which is why antibody proteins were created by random chance or natural selection? If so, why would natural selection bring about disease, viruses, and death when natural selection is supposedly intended to move life forward, not to destroy life. How does evolution justify the creation of all these different types of amino acids that form the different types of proteins if there isn't a need for them at this stage in the development of life? For example, DNA polymerase III has a donut shape specifically designed so that it can form a ring around DNA. Well, according to the evolution timeline, DNA hasn't been created yet. RNA might be on the scene depending on what source you reference, but RNA is a single strand structure, and DNA is a double-strand structure. How does natural selection justify the unique and specific shape of the DNA polymerase III if DNA is not in existence yet? Now that we are about to see what DNA and RNA do for life, here is a little teaser, or some may call it a mystery, for you to ponder on. In a simple explanation of the relationship between proteins, DNA, and RNA. DNA stores the instructions needed to produce proteins. RNA delivers these instructions to the protein manufacturing machine that produces the protein. And in turn, the protein packages the DNA into chromosomes. Think about that, DNA gives the instructions needed to make proteins to RNA, which delivers those instructions to the manufacturing machine. So how does protein randomly form first? Or how do all three form independently when they are dependent on each other?

CHAPTER 6
DNA and RNA

DNA

The National Human Genome Research Institute defines DNA as follows.

DNA (deoxyribonucleic acid) is made of chemical building blocks called nucleotides. These building blocks are made of three parts: a phosphate group, a sugar group, and one of four types of nitrogen bases. To form a strand of DNA, nucleotides are linked into chains, with the phosphate and sugar groups alternating. The four types of nitrogen bases found in nucleotides are adenine (A), thymine (T), guanine (G), and cytosine (C). The order, or sequence, of these bases, determines what biological instructions are contained in a strand of DNA. For example, the sequence ATCGTT might instruct for blue eyes, while ATCGCT might instruct for brown. DNA consists of two strands spiraled around each other in a double helix pattern.

"The complete DNA molecule or genome, for a human contains about 3 billion bases and about 70,000 genes on 23 pairs of chromosomes . . . DNA contains the instructions needed for an organism to develop, survive, and reproduce. To carry out these functions, DNA sequences must be converted into messages that can be used to produce proteins, which are the complex molecules that do most of the work in our bodies . . . This (Double Helix) shape - which looks much like a twisted ladder - gives DNA the power to pass along biological instructions with great precision. In addition, when proteins are being made, the double helix unwinds to allow a single strand of DNA to serve as a template. This template strand is then transcribed into mRNA, which is a molecule that conveys vital instructions to the cell's protein-making machinery." [36]

The significance of DNA is very high. The gene's sequence is like language that instructs a cell to manufacture a particular protein. An intermediate language, encoded in the sequence of Ribonucleic Acid (RNA), translates a gene's message into a protein's amino acid sequence. It is the protein that determines the trait. This is called the central dogma of life. [37]

Natural selection and evolution have created this extremely complex molecule known as DNA? Seriously? Remember the odds of 10 to the 164th for one smaller than average protein chain. Now consider the complexity of DNA versus protein. What are the odds of DNA being created by chance? How does natural selection explain that when a protein is being made, DNA needs to unwind to allow a single strand of DNA to serve as the template? That template is transcribed into mRNA (messenger RNA), which delivers the instructions to the cell's "protein-making machinery." That is such a complex process that must be absolutely perfect! Wouldn't natural selection want simple over extremely complicated? The more complex, the greater chance of error. The greater chance of error, the greater chance of death. I would think natural selection would favor the acronym "K.I.S" (Keep It Simple). If Evolutionists are correct and life has evolved by natural

---
[36] https://www.genome.gov/about-genomics/fact-heets/Deoxyribonucleic-Acid-Fact-Sheet
[37] https://www.bioinformatics.org/tutorial/1-1.html

selection, then natural selection has gone out of its way to increase the complexity of life at every branch on the evolutionary tree of life, as you will see.

A chromosome is made of two DNA strands spiraled around each other in a double helix pattern. Every life form has its own unique DNA and unique genetic code. My DNA is different from your DNA and every other life form's DNA. How can we possibly believe that trillions of different combinations of DNA needed to represent every life form were created by random chance? How does random chance explain the spontaneous formation of DNA containing genetic code when at this point, there was no life. No genetic code needed to be passed on. It gets better when DNA was just spontaneously forming; how did that DNA strand know when it had captured all the genetic code needed? How did it know how long or short to be? How does a sugar group, a phosphate group, and one of four nitrogen bases being assembled act like a computer hard drive and actually record and store the genetic code for the life form it represents? And why did DNA decide to make an identical strand (DNA has two identical strands spiraled around each other in a double helix pattern) and RNA only needs one strand? How did DNA decide to change shape from a circular strand in bacteria to a linear shape found in plants, animals, and humans?

Let us pause for a second and understand, just since the formation of Earth, everything that has had to have happened by random chance in order for life to have formed. Random chance has had to first create and identify 20 amino acids that will be found in the genetic code and used as the building blocks of proteins for the creation of life; out of the more than 500 total amino acids that random chance has created thus far. And we have learned that the odds of random chance creating just one correctly sequenced peptide, according to Evolutionist, is 1 in 10 to the 40th power. One correctly sequenced smaller than average protein made of 150 amino acids is 1 in 10 to the 164th power. The last probability that we have is one protein made of 300 amino acids is 1 in 10 to the 390th power.

Can you honestly look at how complex the simplest atoms, molecules, and building blocks of life are and rubber stamp it as Sure, I believe it all happened by random chance in some early Earth primordial soup. There is absolutely nothing random about it.

Continuing on, we see that when DNA is discussed, RNA is usually part of that discussion. That is because DNA and RNA go hand in hand, and the only difference between these two molecules is that one starts with a D and the other starts with an R. Ahh, I hope I made you say to yourself, wait, what? I am just kidding around with that statement. Yes, there are some very important differences between DNA and RNA. We just looked at DNA; now, we will look at RNA.

RNA

The National Human Genome Research Institute defines RNA as follows.

RNA (Ribonucleic acid) is a molecule similar to DNA. Unlike DNA, RNA is single-stranded. An RNA strand has a backbone made of alternating sugar (ribose) and phosphate groups. Attached to each sugar is one of four bases--adenine (A), uracil (U), cytosine (C), or guanine (G). RNA is similar to DNA, so what is the primary role of RNA? Its principal role is to act as a messenger carrying instructions from DNA to produce proteins to the protein-making machine inside the cell.

((Ribonucleic acid, or RNA, is one of the three major biological macromolecules that are essential for all known forms of life (along with DNA and proteins). A central tenet of molecular biology states that the flow of genetic information in a cell is from DNA through RNA to proteins: "DNA makes RNA makes protein". . . We now know that RNA can also act as enzymes (called ribozymes) to speed chemical reactions . . . RNA also plays an important role in regulating cellular processes–from cell division, differentiation, and growth to cell aging and death)). [38]

RNA is an amazing molecule that is essential to life. Yet another building block of life. Subatomic particles, amino acids, proteins, DNA, and now RNA are essential for life's existence. Not only are all these molecules needed for there to be life, but they also all have vital roles and responsibilities to perform in conjunction with each other. These molecules are useless by themselves, but together are the reason life is possible. Yet, according to evolutionists, these molecules formed by random chance and somehow bumped into each other and decided to become a team and create the foundation for life. All of this random chance and spontaneous formation of DNA and RNA molecules had to happen in the same place; otherwise, the evolutionists are up a creek without a paddle, so to speak.

Let's see what scientists have to say about the creation of DNA and RNA now that we know their vital roles and responsibilities. Scientists have thoughts on the "evolution of" RNA and DNA, but when it comes to how these molecules came to be, the scientists that advocate evolution are, you

---

[38] https://www.rnasociety.org/what-is-rna

guessed it. Spontaneous formation is the answer. Life isn't possible without them, so these molecules just happened, as discussed in the following two articles.

"One theory is that RNA, a close relative of DNA, was the first genetic molecule to arise around 4 billion years ago, but in a primitive form that later evolved into the RNA and DNA molecules that we have in life today. Today, genetic information is stored in DNA. RNA is created from DNA to put that information into action. RNA can direct the creation of proteins and perform other essential functions of life that DNA can't do. RNA's versatility is one reason that scientists think this polymer came first, with DNA evolving later as a better way to store genetic information for the long haul. But like DNA, scientists theorize that RNA could also be a product of evolution."

In the first sentence of this article, it is stated that RNA came first. The third sentence states RNA is created from DNA. Even if it was a primitive form of RNA, it was still RNA. Now, this RNA is giving DNA control over its existence. How does that line up with the claims of random chance or evolution? That has got to have you scratching your head a bit and asking, What? Must be a misprint. Nope! The article double downs on RNA, coming first in the fifth sentence. How is this supposed to be taken seriously or make any sense? The article continues.

(Chemists at the Georgia Institute of Technology have shown how molecules that *may have been* present on early Earth can self-assemble into structures that *could represent* a starting point of RNA. *The spontaneous formation of RNA building blocks is seen as a crucial step in the origin of life, but one that scientists have struggled with for decades.* "In our study, we demonstrate a reaction that we see as important for the formation of the earliest *RNA-like molecules*," said Nicholas Hud, professor of Chemistry and Biochemistry at Georgia Tech, where he's also the director of the Center for Chemical Evolution.) [39]

"May have been," "Could represent" and spontaneous formation of RNA; seen as a crucial step in the origin of life, but one that scientists have struggled with for decades = they have no answer.

Here in the second article, the RNA World hypothesis is addressed.

"If Earth spontaneously generated life, then the first biological molecule must have arisen, well, spontaneously, from chemical activity. This differs from other biological processes, which function with an assist from enzymes. Enzymes are catalysts that speed up chemical reactions. Without enzymes, we would not be able to produce energy, fight off bacterial invaders, or create new genes." It is interesting that this article starts off with a big unknown of "if life spontaneously formed" and then seems to use that unknown as supporting evidence that RNA must have also spontaneously formed. Enzymes are proteins, and proteins are the basis for life, as the article states in this quote. "In cells, RNA acts as a time-sensitive copy of a DNA genome. Once RNA is copied off of DNA, ribosomes use it as instructions to produce proteins. Without proteins, life is impossible. But the ribosome is itself a mix of protein and RNA. This creates a real chicken-and-the-egg conundrum." The article then states, "Szostak's team noticed that ANA, RNA, and DNA nucleotides could arise from the same precursor – indicating that all three could originate from the same ingredients that existed in primordial Earth . . . how arabinonucleic acids

---

[39] https://news.gatech.edu/news/2013/12/23/new-study-brings-scientists-closer-origin-rna

(ANA) would significantly increase the rate of RNA synthesis and stability in the environment of a primordial Earth."[40]

But when I research arabinonucleic acids (ANA), I find articles like these claiming ANA are artificial and synthesized.

ANA refers to the *artificial* genetic system, arabinose nucleic acid. [41]

Inversion of the configuration of the C2' position of RNA leads to a very unique nucleic acid structure: arabinonucleic acid (ANA). [42]

Arabinonucleic acid (ANA), the 2'-epimer of RNA, was synthesized from arabinonucleoside building blocks by conventional solid-phase phosphoramidite synthesis. [43]

"Arabinonucleic acid (ANA), the 2' -stereoisomer of RJ'J'A, was synthesized via the solid-phase synthesis in order to test for its ability to associate with single and double-stranded nucleic acids (DNA and RNA)" file:///C:/Users/spand/Downloads/Noronha-1999.pdf

Did random chance spontaneously synthesize ANA just like RNA, according to the massivesci.com article?

So many unknowns, yet evolution is considered a fact and mandatory teaching to children to answer the question, how did life begin. This is shameful.

---

[40] https://massivesci.com/articles/origin-life-rna-world-self-reproducing-ana-dna/
[41] https://www.sciencedirect.com/science/article/pii/S2666246921000124
[42] https://www.tandfonline.com/doi/abs/10.1081/NCN-100002317
[43] https://pubmed.ncbi.nlm.nih.gov/10852702/

Chapter 7

The Single Cell

Amino acids, proteins, DNA, and RNA are all vital and fundamental components of life that cannot exist and perform their functions on their own. Without any of these, life would not be possible as we know it. We know that chains of amino acids folded into very specific arrangements make different proteins and that DNA molecules store the genetic code for the life form that it is a part of. We also know that RNA molecules are the messengers that copy the instructions from the DNA and deliver them to the ribosomes and the protein-making machine. Where is this protein-making machine, and where do all these molecules interact with each other? All of this happens in a "Cell." In biology, a cell is the smallest unit that can live on its own and makes up all living organisms and the body's tissues.

What are the components of a cell? What makes the cell special or vital to the continuation of life? "A cell consists of three parts: the cell membrane, the nucleus, and, between the two, the cytoplasm. Within the cytoplasm lie intricate arrangements of fine fibers and hundreds or even thousands of minuscule but distinct structures called organelles."[44]

This is the basic structure of a cell which we will use as the foundation from which to build our knowledge. Now to quote

---

[44] https://training.seer.cancer.gov/anatomy/cells_tissues_membranes/cells/structure.html.

from; Origin: Probability of a Single Protein forming by chance, published April 7, 2017). [45]

At time 6:35, Paul Nelson, Philosopher of Biology, Biola, University, is quoted saying, "Now suppose against all odds, chemical evolution produced our single function protein, would we have life? No. We'd have one protein, just a lifeless arrangement of amino acids. The simplest living cell we know has more than 300 different proteins. But proteins are only a part of the story when you consider any actual cell. Rember, you're going to have carbohydrates, complex sugars, nucleic acids, DNA & RNA, lipids, and a whole variety of different chemicals which jointly constitute the living state. Those bits and pieces all have to be brought into the same microenvironment at the same moment in time."

Ann Gauger, Developmental Biologist, Biologic Institute, is quoted at time 8:20 saying, "From my understanding of what it takes to make a cell, it has to happen all at once. You can't do it one bit at a time because everything works together in a causal loop. The higher level of organization transcends the pieces. The spatial organization in the cell requires that molecules end up in the right place at the right time."

Yet with the understanding that the creation of a cell must happen all at once, Evolutionists still try and claim that life happened over many, many small steps. If the NewScientist article is true, and the nucleus formed first, as quoted here from the article. "Fatty molecules coated the iron-sulfur froth and spontaneously formed cell-like bubbles. Some of these bubbles would have enclosed self-replicating sets of molecules – the first organic cells." Again, another article claims spontaneous formation is how this process happens. If cell-like bubbles did spontaneously form, wouldn't these cell-

---

[45] https://m.youtube.com/watch?v=W1_KEVaCyaA

like bubbles be filled with iron-sulfur froth? Is that good for the "self-replicating sets of molecules"? How then did evolution replace the iron-sulfur froth and evolve cytoplasm and a cell membrane to contain it all? Also, if this is all true regarding evolution, why have cells stopped forming in this manner and can now only form by the process of cell division? Common sense tells us that this is not the way life formed. Another thing that common sense tells us is that the odds of random chance bringing together all the individual molecules that make up a cell at just the right place and right time and being able to repeat this process billions and trillions of times is, well, just plain unbelievable.

There are two types of cells. Eukaryotic and prokaryotic. This is a brief excerpt from the following article describing the differences between the two cell types [46]."All life on Earth consists of either eukaryotic cells or prokaryotic cells. Prokaryotes were the first form of life. Bacteria and Archaea are the two types of prokaryotes. Prokaryotes are unicellular organisms that lack membrane-bound structures, the most noteworthy of which is the nucleus. Prokaryotes do have distinct cellular regions. In prokaryotic cells, DNA bundles together in a region called the nucleoid. Eukaryotes are organisms whose cells have a nucleus and other organelles enclosed by a plasma membrane. Organelles are internal structures responsible for a variety of functions, such as energy production and protein synthesis. Animals, plants, fungi, algae, and protozoans are all eukaryotes. Scientists believe that eukaryotes evolved from prokaryotes around 2.7 billion years ago."

---

[46]https://www.technologynetworks.com/cell-science/articles/prokaryotes-vs-eukaryotes-what-are-the-key-differences-336095.

Does this evolutionary timetable pass the common-sense test? From the singularity to the formation of our planet Earth and solar system, 9.3 billion years. From the formation of Earth to the first life form, 1 billion years. 800,000 years to evolve a cell with membrane-bound structures, which was approximately 2.7 billion years ago. That means that in less than 2.7 billion years, evolution has evolved a one-celled life form into the "5.3 million to 1 trillion species currently living on Earth." [47] and that does not include the species that evolved and have gone extinct, like the dinosaurs and many others. Wow, is that a lot of evolution in a very short time. Keep this in mind as we move along. But before I get ahead of myself, here are some actual facts on the Nucleus, Cytoplasm, and Cell Membrane.

The nucleus controls and regulates the activities of the cell (e.g., growth and metabolism). "The nucleus is one of the most obvious parts of the cell when you look at a picture of the cell. It's in the middle of the cell, and the nucleus contains all of the cell's chromosomes, which encode the genetic material. . .Sometimes things like RNA need to traffic between the nucleus and the cytoplasm, and so there are pores in this nuclear membrane that allow molecules to go in and out of the nucleus. It used to be thought that the nuclear membrane only allowed molecules to go out, but now it's realized that there is an active process also for bringing molecules into the nucleus." [48]

The nucleus is contained by a membrane, and between this membrane and the actual cell membrane is the cytoplasm, which houses all the other organelles that are a part of that cell. "Cytoplasm is a thick solution that fills each cell and is

---

[47] https://phys.org/news/2019-04-species-earth-simple-hard.html
[48] https://www.genome.gov/genetics-glossary/Nucleus

enclosed by the cell membrane. It is mainly composed of water, salts, and proteins. In eukaryotic cells, the cytoplasm includes all of the material inside the cell and outside of the nucleus. All of the organelles in eukaryotic cells, such as the nucleus, endoplasmic reticulum, and mitochondria, are located in the cytoplasm. The portion of the cytoplasm that is not contained in the organelles is called the cytosol. Although cytoplasm may appear to have no form or structure, it is actually highly organized. A framework of protein scaffolds called the cytoskeleton provides the cytoplasm and the cell with their structure."[49]

And what keeps all the cytoplasm contained is the cell membrane. "The plasma membrane, or the cell membrane, provides protection for a cell. It also provides a fixed environment inside the cell, and that membrane has several different functions. One is to transport nutrients into the cell and transport toxic substances out of the cell. Another is that the membrane of the cell, which would be the plasma membrane, will have proteins on it that interact with other cells. Those proteins can be glycoproteins, meaning there's a sugar and a protein moiety, or they could be lipid proteins, meaning that there's a fat and a protein. And those proteins which stick outside of the plasma membrane will allow for one cell to interact with another cell. The cell membrane also provides some structural support for a cell."[50]

What makes up the cell membrane? "With few exceptions, cellular membranes — including plasma membranes and internal membranes — are made of glycerophospholipids, molecules composed of glycerol, a phosphate group, and two fatty acid chains. Glycerol is a three-carbon molecule that

---

[49] https://www.nature.com/scitable/definition/cytoplasm-280/.
[50] https://www.genome.gov/genetics-glossary/Cell-Membrane

functions as the backbone of these membrane lipids. Within an individual glycerophospholipid, fatty acids are attached to the first and second carbons, and the phosphate group is attached to the third carbon of the glycerol backbone. Variable head groups are attached to the phosphate. Space-filling models of these molecules reveal their cylindrical shape, a geometry that allows glycerophospholipids to align side-by-side to form broadsheets."[51]

Time for a quick evolution check. With just looking at how complex the three basic components of a cell are and what those components are made of, ask yourself if, deep down in your soul, you honestly think that random chance happened to have formed amino acids and then correctly sequenced those amino acids into all the unique combinations to form the life-supporting proteins. Then evolution determined that creating a cell was the best way to package DNA, RNA, and proteins and then encapsulate the DNA in the nucleus of that cell protected by a membrane. DNA creates RNA to deliver the instructions to make proteins, so evolution had to create pores in the membrane of the nucleus that would allow the RNA to move outside of the nucleus. Next, evolution figured that it would gather water, salts, and proteins, which creates cytoplasm to protect the nucleus and the organelles outside of the nucleus. Evolution also created a highly organized framework of protein scaffolds called the cytoskeleton to provide the cytoplasm and the cell with their structure. Then evolution realized that the cytoplasm needed to be contained and that the cell needed protection. Evolution solved this issue by arranging glycerophospholipids, molecules composed of glycerol, a phosphate group, and two fatty acid chains that, as quoted earlier, form the plasma and cellular membrane. Wow,

---

[51] https://www.nature.com/scitable/topicpage/cell-membranes-14052567/

just talking about and describing the arrangement of a cell is a complex task, but supposedly it is a no-brainer and an easy task for evolution. Hopefully, you are starting to see some of the gaps that evolutionists just kind of skip over with terms like "must have happened" or "it's still unknown" and now are forming some of your own questions you would like to have answered by evolutionists. If not, that is ok. There are still many more evolutionary claims to evaluate with common-sense questions. And with that, let's take our knowledge of a single cell and ask evolution what comes next?

With all we now know about the cell, the next step is to learn about single-celled life forms, which are classified as "Microorganisms." Wikipedia defines microorganisms as "A microorganism, or microbe is an organism of microscopic size, which may exist in its single-celled form or as a colony of cells." Microorganisms (microbes) are considered the first forms of single-celled life forms.

CHAPTER 8

Microorganisms

The following article begins to describe the evolution of microorganisms. [52] In this article, Michael Marshall, the author, claims that scientists are in agreement that "at some point far back in time, a common ancestor gave rise to two main groups of life; bacteria and archaea. *How this happened, when, and in what order the different groups split, is still uncertain.*" Further in this article, he states that viruses were present 3 billion years ago, which will be important here shortly. When an article claims what happened as fact by describing that process as "How, and what order is still uncertain." Are we supposed to believe these claims as fact?

Here is another article describing the "common ancestor" and the descendants of that ancestor. [53] "According to the evidence, all three domains of life share a common ancestor that probably existed more than 3 billion years ago (bya). Two lines of descent emerged from this ancestor. One line produced modern-day Bacteria. The other gave rise to a common ancestor (~2 bya) of both the Archaea and the Eukarya." Another article makes its claim based on something that "probably existed."

Both articles, as well as many, many other articles on evolution, claim that evolution created bacteria, the first

---

[52] www.newscientist.com/article/dn17453-timeline-the-evolution-of-life/amp/
[53] http://bcs.whfreeman.com/webpub/biology/sadavalife9e/animated%20tutorials/life9e_2601_script.html

single-celled life form and that *all* future life evolved from bacteria, but not all bacteria evolved into more advanced life forms. If evolution and natural selection (the process whereby organisms better adapted to their environment tend to survive and produce more offspring) are moving life forward, why is the original life form, bacteria thriving along with the new life form? And if evolution had to move life forward with another life form because bacteria was not suited for its environment, then why are bacteria essential to maintaining the intricate functions of modern-day life? You might be asking what bacteria's environment was? That is a very good common-sense question, and it turns out that bacteria's environment was and is just about every environment on Planet Earth. The following two articles describe how well bacteria flourishes in its environment, Planet Earth.

"Bacteria are ubiquitous, living in every possible habitat, including soil, underwater, deep in Earth's crust, and even extreme environments like acidic hot springs and radioactive waste.... They are abundant in lakes and oceans, arctic ice, and. [54] They provide the nutrients needed to sustain life by converting dissolved compounds, such as hydrogen sulfide and methane, to energy. They live on and in plants and animals. Most do not cause diseases, are beneficial to their environments and are essential for life. The soil is a rich source of bacteria, and a few grams contain around a thousand million of them. They are all essential to soil ecology, breaking down toxic waste, and recycling nutrients. They are even found in the atmosphere, and one cubic meter of air holds around one hundred million bacterial cells. <u>The oceans and seas harbor around 3×10^{26} bacteria</u> which provide up to 50%

---

[54] geothermal springs.

of the oxygen humans breathe. Only around 2% of bacterial species have been fully studied."[55]

"There are five nonillion bacteria in the Earth's ecosystem, including those found in living beings. That is 5 x 10 on 30th power."[56] Yes, "nonillion" is a real number; I had to look it up. There are 5,000,000,000,000,000,000,000,000,000,000 bacteria in Earth's ecosystem, but it wasn't adapted to its environment according to evolutionists. Common sense is screaming at me that this just doesn't make sense. Now we look into what has evolved from this common ancestor along the two lines of descent.

Kind of makes you wonder why evolution was needed if bacteria conquered the planet. From this "common ancestor" and the two lines of descent, Scientists agree that these early microbes can be divided into six major types: bacteria, archaea, fungi, protozoa (animal-like protists), algae, and viruses but are not sure how this division happened. Here is a quick introduction of the six major types from the website [57]

Bacteria are microscopic, single-celled organisms that have no nucleus and a cell wall made of peptidoglycan. Bacteria are the direct descendants of the first organisms that lived on Earth, with fossil evidence going back about 3.5 billion years. Bacteria are a huge and diverse group. Its members have many shapes, sizes, and functions, and they live in just about every environment on the planet.

Archaea are microscopic, single-celled organisms that have no nucleus and an outer membrane containing unique lipids. Archaea are surrounded by a membrane made up of a type of lipid that isn't found in any other organism. Most archaea also have a cell wall, but theirs is very different from

---

[55] https://en.wikipedia.org/wiki/Bacteria
[56] https://www.worldatlas.com/how-much-bacteria-is-on-earth.html
[57] https://learn.genetics.utah.edu/content/microbiome/intro/.

the peptidoglycan cell wall of bacteria. Archaea are best known for living in extreme environments, but they also live in non-extreme environments, including the human gut and skin.

Fungi are single-celled or multicellular organisms with nuclei and with cell walls made of chitin. They also have membrane-wrapped organelles, including mitochondria. Fungi are important decomposers in most ecosystems. Their long, fibrous cells can penetrate plants and animals, breaking them down and extracting nutrients. Several species of fungi, mostly yeasts, live harmlessly on the human body.

Protists are single-celled or multicellular, microscopic organisms with cell nuclei that aren't plants, animals, or fungi. Protists are a category of left-overs and oddballs that don't fit into other groups, and taxonomists are continually reorganizing them. Several parasitic protists can cause deadly diseases, including malaria, amoebic dysentery, and giardia. But the human body is also home to beneficial and neutral protists.

"Green algae" (a general term), microscopic plants that live as single cells (sometimes with flagella) or long fibers. Microscopic plants generally do not live in or on the human body, but they are important food sources for animals in freshwater and saltwater ecosystems. They also release oxygen, which is essential for animal life.

And finally, viruses. Viruses are microscopic particles made of nucleic acids, proteins, and sometimes lipids. Viruses can't reproduce on their own. Instead, they reproduce by infecting other cells and hijacking their host's cellular machinery. Viruses are specialized to infect a certain host and often a specific cell type within that host. Viruses infect plants, animals, bacteria, or archaea. In our bodies, viruses infect not only our cells but also other microbes that live in our bodies.

And where did viruses evolve from? Here is an honest answer. [58] *"To date, no clear explanation for the origin(s) of viruses exists.* Viruses may have arisen from mobile genetic elements that gained the ability to move between cells. They may be descendants of previously free-living organisms that adapted a parasitic replication strategy. Perhaps viruses existed before and led to the evolution of cellular life. Continuing studies may provide us with clearer answers. *Or future studies may reveal that the answer is even murkier than it now appears."* Why would evolution and natural selection promote viruses, meant to destroy life, not move life forward? Also, why can the other five types of life all reproduce on their own, except for viruses? Just more big unanswered questions from evolutionists.

Now that we understand the six major types of microbes let's look back at the Prokaryotes, the first single-celled life form that came to be approximately 3.5 billion years ago. Bacteria was the first prokaryotic life form, and archaea evolved from bacteria a short time later.

"Archaeal cells have unique properties separating them from the other two domains, Bacteria, and Eukaryota. Archaea are further divided into multiple recognized phyla. Current knowledge on genetic diversity is fragmentary, and the total number of archaeal species cannot be estimated with any accuracy. Estimates of the number of phyla range from 18 to 23, of which only 8 have representatives that have been cultured and studied directly." [59]

---

[58] https://www.nature.com/scitable/topicpage/the-origins-of-viruses-14398218/
[59] https://en.wikipedia.org/wiki/Archaea

Now, archaea are pretty amazing single-celled organisms, as the excerpt from this article reports. "Archaea today have a wide variety of unique metabolisms that allow them to live in the most inhospitable places on Earth. Archaea can eat iron, sulfur, carbon dioxide, hydrogen, ammonia, uranium, and all sorts of toxic compounds, and from this consumption, they can produce methane, hydrogen sulfide gas, iron, or sulfur."[60]

If archaea evolved from bacteria, why move to such extreme environments? How does an organism evolve to being able to survive in deep-ocean hydrothermal vents that can reach 236 degrees Fahrenheit? Wouldn't generation after generation of archaea just get cooked and destroyed? Wouldn't the smart thing for natural selection be to avoid those vents? And if the first life form, bacteria, fed on naturally occurring amino acids, how does evolution explain the ability to consume toxic chemicals for food when there was not a need to? The more likely answer is that these microorganisms were created for a purpose. Bacteria, archaea, and fungi have been able to feed on and break down nuclear waste. That is not an evolved trait, which are microorganisms that were created with a purpose.

The evolutionary tree of life order shows Bacteria first, then the Archaea branch, and the last branch being the Eukaryotes. Now when you research archaea, you will find contradictions like this one in many articles. "It has been proposed that the archaea evolved from gram-positive bacteria..." and in the same article separated by four sentences, "The archaeal lineage may be the most ancient that exists on earth."[61]

---

[60] https://www.independent.com/2010/02/05/archaea-third-domain-life/.
[61] https://bio.libretexts.org/Bookshelves/Introductory_and_General_Biology/.

Just so you did not miss that. The article proposes that archaea evolved from bacteria and, in the same article, suggests that archaea could be the oldest life form on Earth. How is that possible? Are scientists in agreement with the temperature of early Earth? Nope. They don't know if it was hot or cold. Well, they must be in agreement that RNA was first and then evolved DNA. Oops, wrong again. There are a plethora of articles suggesting DNA was first. And another plethora of articles claiming RNA was first. It seems scientists do not appear to agree on the how or why of any critical evolutionary step, but they agree that it must have occurred, kinda like my hypothetical argument that "Aliens" populated the earth. I am not sure how they did it, but it must have happened because here we are, and it supports my hypothesis. I hope you are looking at these articles and seeing all the unknowns and contradictions that are ignored, brushed over, or sometimes stated in contradiction to the so-called facts of the article. Now let us continue with prokaryotes.

Early Earth was anaerobic

Life began on Earth as anaerobic, no oxygen, and for approximately one billion years, bacteria and archaea survived and flourished in pretty much every environment on Earth. "The first bacteria to appear on Earth, some 3.8 billion years ago, had no choice but to be anaerobic because the atmosphere at that time contained no oxygen." [62]

"Bacteria have been the very first organisms to live on Earth. They made their appearance 3 billion years ago in the waters of the first oceans. At first, there were only anaerobic

---

[62] https://thebrain.mcgill.ca/flash/i/i_05/i_05_cl/i_05_cl_her/i_05_cl_her.html.

heterotrophic bacteria (the primordial atmosphere was virtually oxygen-free)."[63]

In anaerobic respiration, "Each oxidant produces a different waste product, such as nitrite, succinate, sulfide, methane, and acetate. Anaerobic respiration is correspondingly less efficient than aerobic respiration. In the absence of oxygen, not all of the carbon-carbon bonds in glucose can be broken to release energy. A great deal of extractable energy is left in the waste products. Anaerobic respiration generally occurs in prokaryotes in environments that do not contain oxygen."[64]

Evolution tells us that life began in the oceans of planet Earth. A planet with an atmosphere that was devoid of oxygen. These first life forms were anaerobic prokaryotes, bacteria followed by archaea, which flourished in every environment on Earth, from mild to extreme. Now evolutionists tell us that "evolution (evolution is the change in the characteristics of a species over several generations and relies on the process of natural selection) and natural selection (the process whereby organisms better adapted to their environment tend to survive and produce more offspring) move life forward." If this is, in fact, the case, how are bacteria and archaea not extremely well suited for every environment on Earth? Why is there a need for evolution, and what needs to be improved?

Why did the next evolutionary milestone have the possibility of nearly destroying all of life? You ask, what event nearly killed off life on Earth? It was the arrival of oxygen in the atmosphere, which is toxic to anaerobic life.

---

[63] https://www.eniscuola.net/en/argomento/bacteria/bacteria-knowledge/the-first-organisms/
[64] https://en.wikipedia.org/wiki/Cellular_waste_product

CHAPTER 9

Great Oxidation Event

Imagine that, the next major evolutionary milestone, and scientists are once again debating the "Great Oxidation Event" and cyanobacteria, as seen in the following article. The second article highlights that scientists do not know how oxygen first accumulated on Earth. The third article explains the extinction of many anaerobic species due to the sudden build-up of oxygen in the atmosphere.

"2.4 billion years ago, the "great oxidation event." Supposedly, the poisonous waste produced by photosynthetic cyanobacteria – oxygen – starts to build up in the atmosphere. Recently, though, some researchers have challenged this idea. They think cyanobacteria only evolved later. Yet others think that cyanobacteria began pumping out oxygen as early as 2.1 billion years ago but that oxygen began to accumulate only due to some other factor, possibly a decline in methane-producing bacteria. Methane reacts with oxygen, removing it from the atmosphere, so fewer methane-belching bacteria would allow oxygen to build up." [65]

"It is thus ironic that when and how O2 first accumulated on Earth remains a major puzzle." [66]

"The Great Oxidation Event (GOE), also called the Great Oxygenation Event, the Oxygen Catastrophe, and the Oxygen

---

[65] https://www.newscientist.com/article/dn17453-timeline-the-evolution-of-life/
[66] https://www.sciencedirect.com/science/article/pii/S0960982209011890

Crisis, was a time interval when the Earth's atmosphere and the shallow ocean first experienced a rise in the amount of oxygen. This occurred approximately 2.4–2.0 Ga (billion years ago) during the Paleoproterozoic era. Geological, isotopic, and chemical evidence suggests that biologically-produced molecular oxygen (dioxygen, O2) started to accumulate in Earth's atmosphere and changed it from a weakly reducing atmosphere practically free of oxygen into an oxidizing atmosphere containing abundant oxygen. The sudden injection of toxic oxygen into an anaerobic biosphere caused the extinction of many existing anaerobic species on Earth."[67]

Just when life was at its peak and conquering the planet, evolution nearly destroyed life by the evolving cyanobacteria, which produced oxygen as a waste product, by the process of photosynthesis, which is toxic to anaerobic life. Evolution is supposed to further life, but in this case, evolution appears to be suicidal, and again, evolution maintains the original life form. Evolution has created the first two branches of life, bacteria and archaea. Now that cyanobacteria have evolved and are pumping the atmosphere full of oxygen, evolution evolves the third branch of life, the Eukaryotes. Eukaryotes supposedly evolved from archaea around 2.7 billion years ago in this order, Protists (2 billion years ago), Fungi (1.5 billion years ago), and Algae coming on scene 1.4 – 1.2 billion years ago. This timeline is found in many articles regarding the evolutionary timeline of eukaryotes. However, none of the articles seem concerned about the 700-million-year gap between when eukaryotes evolved from archaea and when the first eukaryotic protists came to be. I wonder why that is?

---

[67] https://en.wikipedia.org/wiki/Great_Oxidation_Event

Why did evolution create photosynthesis? An entirely new way to metabolize energy, which just happened to create oxygen as a waste product. Oxygen is crucial for two significant reasons that I will explain shortly. But first, we must ask the obvious common-sense question, why would evolution mess with life again when it was working great?

Cyanobacteria and Photosynthesis

"By 2.7 billion years ago, a new kind of life had established itself: photosynthetic microbes called cyanobacteria, which were capable of using the Sun'-s energy to convert carbon dioxide and water into food with oxygen gas as a waste product." [68]

"Cyanobacteria are the first organisms known to have produced oxygen. By producing and releasing oxygen as a byproduct of photosynthesis, cyanobacteria are thought to have converted the early oxygen-poor, reducing atmosphere into an oxidizing one, causing the Great Oxidation Event." [69]

"Protists are eukaryotes that first appeared approximately 2 billion years ago with the rise of atmospheric oxygen levels." Early Eukaryotes. (2020, August 15). [70]

Photosynthesis uses light energy to change water and carbon dioxide into oxygen and nutrients called sugars. Sugars are used as food, and oxygen is released into the air. (summary of the main point) [71]

"Photosynthesis evolved 3 billion years ago and released oxygen into the atmosphere. Cellular respiration evolved after that to make use of the oxygen." [72]

---

[68] https://www.amnh.org/explore/videos/earth-and-climate/the-rise-of-oxygen/earth-without-oxygen.
[69] https://en.wikipedia.org/wiki/Cyanobacteria
[70] https://bio.libretexts.org/@go/page/13577.
[71] https://kids.britannica.com/kids/article/photosynthesis/353624.
[72] https://flexbooks.ck12.org/cbook/ck-12-biology-flexbook-2.0/section/5.4/primary/lesson/first-cells-bio/

"Photosynthesis is the only significant solar energy storage process on Earth and is the source of our food and most of our energy resources."[73]

"Across the entirety of life, oxygenic photosynthesis, to our knowledge, evolved only once, providing the metabolic singularity that sustains our planet."[74] And from the same article, "We know embarrassingly little about how oxygenic photosynthesis could have evolved. It cannot be overstated how many evolutionary steps and turns must have occurred to arrive at oxygenic photosynthesizing cyanobacteria from any of their anoxygenic phototrophic bacterial counterparts. In this remarkable transition, a number of multi-subunit complexes, enzymes, and metabolisms would each have to evolve and yet, with our exponentially growing knowledge of the diversity of extant microbial life, we have found almost no clues as to how this incredible process occurred at the molecular level. The crowning achievement during this development was the innovation of the oxygen-evolving complex, which oxidizes water to oxygen, providing the electrons necessary to drive oxygenic photosynthesis. Next, a rewiring of the electron transport would necessitate the evolution of the cytochrome b6f complex and soluble electron carrier (i.e., plastocyanin) in order to connect two photosystems in parallel with one another. Downstream enzymes and metabolisms such as the Calvin–Benson cycle for carbon fixation and aerobic respiration would also have to evolve, as there most likely was no aerobic respiration occurring in the Archean atmosphere devoid of oxygen. Each one of these inventions is a feat in its own, but still, only cyanobacteria would become the bacterial lineage to invent

---

[73] https://www.ncbi.nlm.nih.gov/pmc/articles/PMC2949000/.
[74] https://www.sciencedirect.com/science/article/pii/S0960982215004972.

and commandeer all of them together for what would become the metabolism driving our global carbon cycles."

Amazing that article after article admits that scientists do not know how life went from anoxygenic to oxygenic, yet it is expected to be accepted as fact. One reason I cited this article is because the author is pretty honest about admitting that the scientific community does not have the answers and points out many of the steps that, although, they do not know how they happened, these steps would have needed to happen to support the claim of evolution. We must ask this commonsense question. Honestly, what are the odds of the complexity, microscopic precision, and as the author of the article states, rewiring that had to take place to make oxygenic photosynthesis possible, actually being the result of evolution? Then we must remember that this miracle did not occur once but had to occur billions upon trillions of times. What are those odds? At this point, how can even the most educated experts not start to wonder if maybe, just maybe, life was created by a Creator and that Creator created everything from protons to, at this point in the book, oxygenic photosynthesizing cyanobacteria with a specific purpose and role to fulfill in the circle of life?

Another question is, why wouldn't natural selection consider this new evolutionary process of photosynthesis a positive trait and pass it along to every new life form that evolved after cyanobacteria? Also, evolutionists describe horizontal gene transfer, so why wouldn't evolution transfer photosynthesis horizontally to every present life form at that time in history. All life could get all of its energy needs from sunlight! No grocery stores needed. When you are hungry, just go outside for a sunny walk and come back full and satisfied. Of course, if that happened, we would never know the amazing taste of a perfectly cooked ribeye steak, steamed lobster, or ranch dressing. Oh, the humanity!

"Photosynthesis evolved 3 billion years ago and released oxygen into the atmosphere. Cellular respiration evolved after that to make use of the oxygen."[75]

The first critical point about oxygen:

"Aerobic cellular respiration is the process by which the cells of a living organism break down food and turn it into the energy they need to perform their essential functions. The importance of aerobic respiration in living things cannot be underestimated. Without this process, no living thing would survive." "Aerobic respiration is a series of reactions in which glucose releases energy. Glucose and oxygen are used up, producing carbon dioxide as waste. The cells of humans, animals, and plants go through this process constantly, even when at rest."[76]

The second critical point about oxygen:

Oxygen is toxic and kills anaerobic microorganisms, as stated in this article. "Anaerobic microorganisms lack and don't possess certain enzymes such as catalase, oxidase, superoxide dismutase that are essential for bacteria to survive in the presence of oxygen. Due to the absence of these enzymes, oxygen becomes toxic to anaerobic bacteria."[77]

Now think of this, life in the form of bacteria and archaea was conquering every environment on Earth, feeding and multiplying without oxygen. Oxygen was deadly to these anaerobic life forms. Now, remember evolution (evolution is the change in the characteristics of a species over several generations and relies on the process of natural selection) and

---

[75]https://flexbooks.ck12.org/cbook/ck-12-biology-flexbook-2.0/section/5.4/primary/lesson/first-cells-bio/
[76]https://sciencing.com/importance-aerobic-cellular-respiration-6376108.html
[77] https://www.vedantu.com/question-answer/oxygen-toxic-to-anaerobic-bacteria-class-11-biology-cbse-611b34926c2944748b101fe6

natural selection (the process whereby organisms better adapted to their environment tend to survive and produce more offspring) move life forward.

Evolution then creates a new life that, as a waste product, produces oxygen, toxic and deadly to life. So, the life that was doing great is forced to evolve or die because "evolution" is producing oxygen, which kills the life that evolution had evolved to this point, anaerobic microorganisms. Honestly, think about that. Evolution and natural selection that moves life forward are responsible for creating oxygen, which at that time kills life. A little bit of a conundrum there. And even better, evolutionists do not see an issue with that. Evolution could have actually wiped-out life with this "evolutionary step."

Also, after all that evolution, all the "rewiring," new functions, new roles, new molecules that had to be "invented," and everything else that had to happen to evolve this new way of life, there are still anaerobic bacteria and archaea? Why?

"Although most protists require oxygen (obligate aerobes), there are some that may or must rely on anaerobic metabolism—for example, parasitic forms inhabiting sites without free oxygen and some bottom-dwelling (benthic) ciliates that live in the sulfide zone of certain marine and freshwater sediments."[78]

Hopefully, this is getting your brain fired up, and you are starting to ask some of your own questions, some of which I am sure are even deeper than the ones I have presented. Questions need to be asked, explored, and investigated.

It is time to strap in and hold on to the evolution timeline because dates and the corresponding evolutionary milestones are going to start flying at us like asteroids in the ever-

---

[78] https://www.britannica.com/science/protist/Respiration-and-nutrition

expanding universe. The more complex the evolutionary milestone, the shorter the time needed for said evolution. No more billion-year chunks of time to evolve a cell. Now we only need millions of years or less to make giant evolutionary progress. We will continue to investigate the evolutionary timeline and stop at a few of these giant evolution claims and see how they hold up to some basic common-sense questions. Also, I realize that I often repeat timelines or smaller parts of bigger timelines. I do this because timelines give you a quick mental picture that allows you to ask yourself what sounds reasonable and what sounds like fantasy.

For example, the statement, 9.3 billion years to form our solar system, might sound reasonable. I mean, that is a really long time. Approximately one billion years to form a peptide may not be as reasonable, but one wants to believe it when stated as a fact. Next is a cell, the complexity of each different type and quantity of the individual molecules needed for the complex and perfectly arranged structure in the creation of the cell. And evolution was able to create this miracle in 80% less time than it needed to produce a peptide, which is not even a fully functioning protein. Now we are definitely in the fantasy category. Do you see how the more complex the next step in evolution is, the smaller amount of time is needed to complete this step? So please allow me once more to repeat a few important dates to keep in mind as we move forward.

The universe formed- 13.8 billion years ago, Earth formed – 4.5 billion years ago, first life formed – 3.7 billion years ago, first Single-celled prokaryotes, life forms – 3.5 billion years ago, Eukaryotes evolved from prokaryotes – 2.7 billion years ago. Here we are now in a world full of prokaryotes and eukaryotes (bacteria, archaea, fungi, protists, algae, and viruses).

Earth is now populated with microbes that are divided into six major types: bacteria, archaea, fungi, protozoa, algae, and viruses which evolutionists state evolved into every life form past and present that we have identified here on earth. What did these first life forms eat?

All protozoa are heterotrophic, deriving nutrients from other organisms, either by ingesting them wholly by phagocytosis or taking up dissolved organic matter or microparticles (osmotrophy). [79]

"The fungi had to be eating something. One possibility: bacteria. Researchers have found signs that crusts of bacteria were growing on land as long as 3.2 billion years ago. It's also possible that these ancient fungi lived on the bottom of the estuary, perhaps feasting on underwater mats of algae." [80]

"Researchers have captured images of green alga consuming bacteria." [81]

Evolution (evolution is the change in the characteristics of a species over several generations and relies on the process of natural selection) and natural selection (the process whereby organisms better adapted to their environment tend to survive and produce more offspring) move life forward.

Evolution created Eukaryotes, protists, fungi, and algae to feed on bacteria, the first life form. Well, that's not good for bacteria. Evolution and natural selection do not seem all that concerned about moving life forward for bacteria. Evolution evolves a new life form to kill and feed on another life form. This will be a recurring theme as well. Evolution and natural selection will move life forward by creating new life forms to feed on early life forms or feed on other life forms. Seems like

---

[79] https://en.wikipedia.org/wiki/Protozoa#Feeding
[80] https://www.nytimes.com/2019/05/22/science/fungi-fossils-plants.html
[81] https://www.sciencedaily.com/releases/2013/05/130523143741.htm

for the prey, evolution and natural selection are not moving life forward but trying to end their life form. Yikes!

Chapter 10

SNOWBALL EARTH

Next on the evolutionary timeline, because of the great oxidation event, is the Earth freezes over into the first "Snowball Earth." Do not be surprised that there are many different opinions among scientists as to how many snowball earth events there were, the dates that these events took place, and how long these events lasted. There are even debates about whether a snowball event took place or if it was more of a "Slushball" event. Go figure, yet another example of an evolutionary milestone that scientists cannot agree if it happened, when it happened, or how many times it happened. But evolution is a proven fact. According to evolutionists who support the snowball earth event, it is a very important step in evolution. Many articles make the same claim as this quote from the Wikipedia article that "the most recent snowball episode may have triggered the evolution of multicellularity." That is a pretty big claim to make with so much confusion and debate on everything that is needed to achieve multicellular life.

Here are four articles showing that scientists say anywhere from one to four events took place. Some articles will call them "snowball" events; some just avoid the debate and call them "severe" ice ages. Two articles agree that the event could have happened 2.4 billion years ago, and two articles say it happened between 700 million years and 580 million years ago. These events range from as short as three million years to as long as 58 million years. That is unbelievable. The third article briefly addresses the "slushball" theory but quickly dismisses it because that article promotes the "snowball" theory. If one of these groups of scientists explained the theory they are trying to prove to an audience; I can see that audience walking away saying, wow, pretty amazing. But when you read multiple articles on the subject, you walk away thinking nobody has a clue. They are still just throwing "snowball" ideas against a wall and seeing if anything sticks.

'The Snowball Earth hypothesis proposes that the planet's surface became entirely or nearly entirely frozen during one or more of Earth's icehouse climates. It is believed that this occurred sometime before 650 Mya (million years ago) during the Cryogenian period. . . occurred 2400 to 2100 Mya, may have been triggered by the first appearance of abundant oxygen in the atmosphere, which was known as the "Great Oxidation Event." [82]

"Snowball Earth hypothesis, in geology and climatology, an explanation first proposed by American geobiologist J.L. Kirschvink suggesting that Earth's oceans and land surfaces were covered by ice from the poles to the Equator during at least two extreme cooling events between 2.4 billion and 580 million years ago." [83]

"Snowball Earth episode appears to have come between 700 million and 600 million years ago, when scientists think ice smothered the entire planet, from the poles to the equator — twice in quick succession." "Alternative theories have arisen over the years, including what is called the Slushball theory- In the Slushball scenario, carbon dioxide would start

---

[82] https://en.wikipedia.org/wiki/Snowball_Earth
[83] https://www.britannica.com/science/science

building up very quickly, so the glaciation would be short-lived, and the ice would retreat gradually." [84]

"Scientists believe that three to four severe ice ages, which froze nearly or all of the surface, occurred between 750 million and 580 million years ago." [85]

The Snowball Earth event('s) are very interesting. What might have caused the events is also interesting. In the astronomy.com article, the theory on how this event started is as follows.

"When the Snowball events occurred, the supercontinent Rodinia was in the process of breaking up. The reason why people think there is a connection there is that because the breakup of a supercontinent would increase rainfall in the continental areas, and that would increase the weathering of crustal rocks. The weathering of rocks actually consumes carbon dioxide, so that would lead to less carbon dioxide in the atmosphere and, therefore, a colder climate. As for what actually caused the immediate trigger, attention has focused in recent years on a sequence of very large volcanic eruptions that occurred in what is now the high arctic of Canada. These eruptions occurred around 717 million and 719 million years ago. When you get fire fountains — lava that comes out of one place over a period of weeks or months — you get a strong thermal upwelling in the atmosphere from the heating effect of that lava. These upwellings can loft sulfur aerosols into the stratosphere, where they hang around for a significant amount of time. These sulfur gas particles reflect incoming solar radiation and have a strong cooling effect. Because of the coincidence in timing between these eruptions and the onset

---

[84] https://astronomy.com/news/2019/04/the-story-of-snowball-earth
[85] https://www.livescience.com/64692-snowball-earth.html

of the first and longer of the two Snowball Earths, it's been postulated that that may have been the immediate trigger."[86]

It is funny to think that volcanic eruptions will actually cool the earth. And research on this topic does say that volcanic eruptions will cool the earth, and there is recent data to back this claim up.

"At their most potent, volcanoes inject millions of tons of Sun-blocking particles high into the atmosphere that can cool Earth for nearly 5 years. The most recent, the Philippines' Mount Pinatubo eruption in 1991, caused a temporary 0.5°C drop in global temperatures."[87]

They could cool the Earth at their most potent for nearly five years. The shortest estimate of one of these snowball events is 3 million years, and the longest is 58 million years. That is a monumental difference between cooling the Earth for five years to the snowball event lasting between three and 58 million years. I have not found any articles trying to justify that gap. Just that volcanoes are the trigger for the snowball, oddly enough, later, you will see that scientists claim that volcanic eruptions were also the trigger that ended the snowball event. Well, isn't that special. Could the volcanic eruptions cool the Earth enough to kick off the snowball event in those five years? Researching volcanic eruptions, you will also find articles like this one that say the cooling effect is short-term, but the long-term effect is the warming of the planet. "Over the long term, they belched carbon dioxide from Earth's interior, causing warming. But in the short term, their sulfur gases often react with water to form highly reflective particles called sulfates, triggering spells of global cooling."[88]

---

[86] https://astronomy.com/news/2019/04/the-story-of-snowball-earth
[87] https://www.science.org/content/article/massive-volcanoes-could-cool-earth-more-warming-world
[88] https://www.science.org/content/article/massive-volcanoes-could-cool-earth-more-warming-world

Well, volcanic eruptions do it all, apparently, whether that is to cause short term effects extreme enough to freeze the planet for millions of years or to then warm the planet enough to melt all the ice and then somehow repeat the cycle millions of years later, depending on the article you read. Wow!

How long did these snowball earth events last, and how did life survive?

"The first one lasted 58 million years, and the second one only lasted 5 million to 15 million years. So, we don't know why there is this great disparity in how long the glaciations lasted. And why was it that there was just this short interval between the two? There were only about 10 million years when there was no ice at all, and then suddenly, the planet went back into Snowball Earth. So why two in rapid succession? And why wasn't there a third one or a fourth one? These are new questions that have arisen as a result of our understanding of the timing."[89]

"Scientists contend that at least two Snowball Earth glaciations occurred during the Cryogenian period, roughly 640 and 710 million years ago. Each lasted about 10 million years or so."[90]

". . . which each lasted approximately 10 million years."[91]

". . . suggest that the glacial episode lasted for at least 3 million years."[92]

I am not going to `expand any more here on the length of the snowball events. I mentioned some of the different lengths of snowball events earlier, but I wanted to give you articles showing the major discrepancies. I do want to look at life during the time of the snowball event. Up to this point,

---
[89] https://astronomy.com/news/2019/04/the-story-of-snowball-earth
[90] https://www.giss.nasa.gov/research/features/201508_slushball/
[91] https://www.livescience.com/64692-snowball-earth.html
[92] https://en.wikipedia.org/wiki/Snowball_Earth

according to evolution timelines, there are anaerobic bacteria and archaea that were flourishing, but due to the evolution of cyanobacteria and photosynthesis producing and releasing oxygen, anaerobic bacteria and archaea are in a fight for survival. The oxygen is killing them, and they are about to be a big player in the snowball Earth event. As the Earth was covered in a thick sheet of ice, up to 328 feet thick, that ice now blocks out the sunlight needed for photosynthesis. That means that the cyanobacteria cannot produce oxygen. According to [93] sunlight can penetrate the open ocean to a depth of 100 to 200 meters. That is the ocean that is not covered by 100 meters of ice.

Here again, countless articles claim mass extinction during this event, and countless other articles claim otherwise; here is an excerpt from Wikipedia that represents both sides in one entry. "Some argue that this kind of glaciation would have made life extinct entirely. However, organisms and ecosystems, as far as it can be determined by the fossil record, can determine it, do not appear to have undergone the significant change that would be expected by a mass extinction. With the advent of more precise dating, a phytoplankton extinction event which had been associated with snowball Earth was shown to precede glaciations by 16 million years. Even if life were to cling on in all the ecological refuges listed above, a whole-Earth glaciation would result in a biota with a noticeably different diversity and composition. This change in diversity and composition has not yet been observed—in fact, the organisms which should be most susceptible to climatic variation emerge unscathed from the snowball Earth. One rebuttal to this is the fact that in many of these places where an argument is made against a mass

---

[93] https://phys.org/pdf4677.pdf.

extinction caused by snowball Earth, the Cryogenian fossil record is extraordinarily impoverished."[94] Again, scientists and evolutionists claim both sides of the coin, yet evolution itself is not questionable but taught as a proven fact.

How or what caused the snowball earth events to cease?

'Over 4 to 30 million years, enough CO2 and methane, mainly emitted by volcanoes but also produced by microbes converting organic carbon trapped under the ice into the gas, would accumulate to finally cause enough greenhouse effect to make surface ice melt in the tropics until a band of permanently ice-free land and water developed;[58] this would be darker than the ice, and thus absorb more energy from the Sun—initiating a "positive feedback."'[95]

"Although the team doesn't know for certain what caused it, carbon dioxide emitted by ancient volcanoes may have triggered a greenhouse event, causing the ice sheets to thaw rapidly."[96]

"It seems that the accumulation of millions of years' worth of carbon dioxide emissions from ongoing volcanic activity led to sufficient atmospheric warming to rapidly melt the ice cover. . . widespread explosive underwater volcanism."[97]

Now pretty much all the articles claim to one degree or another that volcanic eruptions emitting massive quantities of carbon dioxide are what lead to the warming of the atmosphere and eventually were able to end the snowball event. I do find it interesting that the build-up of carbon dioxide being belched out by erupting volcanoes over millions of years led to the greenhouse event that ended the snowball event, but at the same time might have triggered the snowball

---

[94] https://en.wikipedia.org/wiki/Snowball_Earth
[95] https://en.wikipedia.org/wiki/Snowball_Earth
[96] https://www.science.org/content/article/ancient-snowball-earth-thawed-out-flash
[97] https://www.realclearscience.com/articles/2016/01/20/how_snowball_earth_came_to_a_fiery_end_109514.html

event. Then there are articles like this one here that state carbon dioxide has never caused detectable global warming.

"While sulfur dioxide released in contemporary volcanic eruptions has occasionally caused detectable global cooling of the lower atmosphere, the carbon dioxide released in contemporary volcanic eruptions has never caused detectable global warming of the atmosphere.... While it has been proposed that intense volcanic release of carbon dioxide in the deep geologic past did cause global warming, and possibly some mass extinctions, this is a topic of scientific debate at present." [98]

It really gets tough trying to follow all the contradictions of evolution against its own claims. At the end of the snowball event, the speed at which it ended is very interesting. The entire planet was covered in a very thick sheet of ice for millions of years, 3 – 58 million, and some articles report that the ice was gone in under one million years. That is extremely fast! Scientists claim that underwater volcanoes and microbes produced "The carbon dioxide levels necessary to thaw Earth have been estimated as being 350 times what they are today, about 13% of the atmosphere." [99]. My question here is that if undersea volcanoes and microbes produced carbon dioxide under the ice sheet, how did the gas get through the ice and into the atmosphere? Some researchers claim that the ice was greater than 100 meters thick, that's 328 feet thick. That is ice that is thicker than a football field stood up vertically from goal line to goal line. That is some seriously thick ice. And a lot of scientists believe this cycle happened multiple times.

---

[98] https://www.usgs.gov/programs/VHP/volcanoes-can-affect-climate
[99] https://en.wikipedia.org/wiki/Snowball_Earth

SNOWBALL EARTH HAS ENDED – 2.15 BILLION YEARS AGO 1ST PHOTOSYNTHESIS

The snowball event has ended, and the Earth is no longer covered in a thick blanket of ice. Life can now get back to evolving, and with that, what surprises does evolution have in store for us now? Well, for starters, every date or discovery we have looked at so far has had multiple articles debating the historical dates or some manner of the evolutionary milestone. So, it should be no surprise that there are also disputes around the date photosynthesis first evolved. We could use some of the earlier dates, which is even worse for the evolutionists and the theory of evolution, but we are going to use the more conservative date that the first "undisputed" evidence of photosynthesis and cyanobacteria in the fossil record was 2.15 billion years ago. Now at 2 billion years ago is when evolutionists claim that the "Eukaryotic" cells first came into being. Funny, that is another way of saying the eukaryotic cells were created. Earlier, I introduced eukaryotic cells when discussing prokaryotic cells. Eukaryotic cells are single-celled life forms with internal organs, referred to as "organelles." The most notable of these organelles is the nucleus. As discussed much earlier, the nucleus is the control center for the cell. Much like the control center for a large cruise ship or container ship is the bridge, and the control center for a commercial airplane is the cockpit.

All the genetic material, chromosomes made up of DNA, for that life form is stored and protected there in the nucleus. At this time along the evolutionary timeline, Eukaryotes branched off from the prokaryotes, and archaea, evolving into protists, fungi, and algae. But don't you worry about any more disputes; oops, I mean, yes, of course, there are many disputes on the timeline that eukaryotes evolved [100] dates the first eukaryotes as Blue-green algae, having come into being 2.5 billion years ago. And this article from Scientific America; [101] claims the "best guesses" for them when the eukaryotes evolved is between 3.5 billion years ago to just under 2 billion years ago. You can also find more "best guesses" if you would like to look, but this is a pretty good range for our argument against evolution and a pretty bad range for the evolutionist argument. I can confidently claim that all scientists can agree that eukaryotes "came into being" after the Earth was formed and before yesterday. Well, I got a kick out of that one. Thank you for letting me have some fun. Now, if we use 3.5 billion years ago for when eukaryotes came into being, that is when prokaryotes came into being, and I haven't read a single article suggesting that they both "evolved" at the same time. That would lead one to believe that a Creator created them at the same time. That seems totally possible.

    A common opinion among evolutionists on just how the eukaryotes evolved from prokaryotes is by symbiosis between two prokaryotes. Many articles describe this as separate prokaryotic cells joining together in a symbiotic union or relationship. How did this joining happen, you ask? Let me answer that. A free-living bacterium was "engulfed" by another prokaryotic cell that was feeding, but the bacterium was not digested but started living in a symbiotic union inside the other cell. I guess that is like saying, when I eat a raw oyster, I do not want to digest the oyster because I am actually trying to form a symbiotic union with the oyster.

---

[100] http://earlyearthcentral.com/early_life_page.html
[101] https://www.scientificamerican.com/article/when-did-eukaryotic-cells/

A common-sense question here is how does the engulfed cell feed itself inside the host cell? Apparently, the engulfed cell fed on the nutrient-rich environment of the host cell. Wouldn't this be considered a parasitic relationship and be bad for the host cell? When my dog gets tapeworms, it is a parasitic relationship, not a symbiotic union/ relationship. The worm is feeding off the nutrient-rich environment of my dog's digestive system, but the tapeworm is robbing my dog of those nutrients. If this relationship goes on without treatment, it can actually kill my dog. It seems like the engulfed cell would be feeding on nutrients that the host cell took in for itself to feed on and would ultimately starve the host cell to death. Then there is the waste matter that the engulfed cell produces. A tapeworm is in the digestive tract of my dog, and therefore the waste matter from the tapeworm processes out along with my dog's waste matter, but the engulfed cell would not be in the "digestive tract" of the host cell. So how does that host cell identify and discard the foreign waste created by the engulfed cell? Also, I have not seen evolutionists try and answer this question.

How does the engulfed cell become the nucleus of the host cell? As we have learned, a nucleus contains the DNA, genetic code, only. The nucleus does not have any other organelles. What happens to the organelles of the engulfed cell if it is now the nucleus? It must move everything but the genetic code outside of its cell membrane into the host cell's cytoplasm. But didn't that just create a cell with two different sets of genetic code? One set is inside the nucleus (engulfed cell), and the second set of genetic codes is outside the nucleus (host cell). What happens to the genetic code of the host cell?

Seriously, evolutionists? One prokaryotic cell engulfed another, resulting in a symbiotic union and the evolution of the eukaryotic cell. This magic formula was somehow shared with millions and trillions of other prokaryotic cells to continue this evolutionary process, but it was not shared with all of the prokaryotic cells because if it was, there would no longer be any prokaryotic cells, only eukaryotic cells. And better yet, evolution is taught as a non-disputable fact.

Still, evolutionists do not seem to be too concerned with any of that and just carry on by saying the engulfed cell evolves into a single lineage, called endosymbiosis (symbiosis in which one of the symbiotic organisms lives inside the other, New Oxford American Dictionary). Also, remember from our discussion of DNA, the DNA is circular in prokaryotic cells, but now in eukaryotic cells, the DNA is linear and protected by the nucleus. They also have mitochondria, and it is in the mitochondria where cellular respiration occurs. They also contain a short loop of DNA that is distinct from the DNA contained in the cell's nucleus. It is from the eukaryotes that, according to evolutionists, multicellular organisms eventually evolve, which we will address at that point in the evolutionary timeline.

Evolutionists describe the different lineages of eukaryotes and claim that around 1.5 billion years ago, they were divided into three groups, which are supposedly the ancestors of modern plants, fungi, and animals. It is around this time when that division actually splits into separate and distinct lineages, and each lineage evolves separately. Some articles state that the order in which these lineages broke off is unknown, and others claim that plants broke off first, followed by fungi, with the animal lineage being the last to break off. Imagine that, yet another step-in evolution that scientists disagree on. But yet, evolution is a fact. I know this seems like a petty point to bring up, but it is important because if they cannot agree or prove the simpler steps, where is their credibility on the more complex issues of when and how unicellular life "evolves" into multicellular life?

CHAPTER 11

Multicellular life

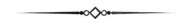

Evolution (evolution is the change in the characteristics of a species over several generations and relies on the process of natural selection) and natural selection (the process whereby organisms better adapted to their environment tend to survive and produce more offspring) move life forward. I am repeating this because I need to ask, why is there a need to "evolve" from unicellular life to multicellular life? Unicellular life is at the top of the food chain. This life has flourished in every imaginable and unimaginable environment on Planet Earth. It survived and multiplied without oxygen, and when evolution evolved cyanobacteria which produced oxygen and threatened to wipe out most of life on Earth, single-celled organisms, according to evolutionists, evolved aerobic capabilities. Evolution serves no purpose here other than to complicate life yet again by now evolving multicellular life.

What exactly is multicellular life, and how do multicellular organisms differ from unicellular (single cell) organisms?

"A colony of single-cell organisms is known as colonial organisms. The difference between a multicellular organism and a colonial organism is that the <u>individual organisms that form a colony or biofilm can, if separated, survive on their own, while cells from a multicellular organism (e.g., liver cells) cannot.</u> Multicellular organisms carry out their life processes through division of labor, and they have specialized cells that have specific functions."[102]

Common-sense asks the question, why can't the cells of a multicellular organism survive on their own if separated? Are the cells of multicellular organisms different than the cells of unicellular organisms?

"Multicellular organisms are composed of more than one cell, with groups of cells differentiating to take on specialized functions. In humans, cells differentiate early in development to become nerve cells, skin cells, muscle cells, blood cells, and other types of cells. One can easily observe the differences in these cells under a microscope. Their structure is related to their function, meaning each type of cell takes on a particular form in order to best serve its purpose. Nerve cells have appendages called dendrites and axons that connect with other nerve cells to move muscles, send signals to glands, or register sensory stimuli. Outer skin cells form flattened stacks that protect the body from the environment. Muscle cells are slender fibers that bundle together for muscle contraction. The cells of multicellular organisms may also look different according to the organelles needed inside of the cell. For example, muscle cells have more mitochondria than most other cells so that they can readily produce energy for movement; cells of the pancreas need to produce many proteins and have more ribosomes and rough endoplasmic reticula to meet this demand. Although all cells have organelles in common, the number and types of organelles present reveal how the cell functions."[103]

It seems that multicellular organisms are carefully created, with individual cells created to perform specific functions for

---
[102] https://www.ck12.org/book/ck-12-biology-advanced-concepts/section/3.24/
[103] https://www.nationalgeographic.org/encyclopedia/unicellular-vs-multicellular/

that organism to survive. Multicellular organisms, from fungi to humans, are created with cells specifically designed to perform a life-sustaining function. If that cell is separated from the organism, that cell will not be able to survive. It will die. Evolutionists understand how monumental this difference between unicellular and multicellular organisms is. When asked about this step in evolution, the answer is, quoting from the New Scientist article, "It is unclear exactly how or why this happens." Imagine that. Again, they are confident that it happened but don't know how or why. Yet they demand that you accept the fact that this did happen because of evolution. Both the how and why are very important and are not because of evolution but point to the work of a Creator.

With evolution moving life forward, what is next for multicellular life?

I think you are going to laugh in disgust at what evolutionists are saying has happened in the articles below. I honestly thought it was a joke because I had never heard of this. And I was especially confused when it was associated with evolution. Then in my research for this book, I discovered many more articles on this subsection of multicellular life. What am I referring to? Alright, enough teasing; some scientists refer to this event as "vacillation," which in the Century Dictionary is defined as; The act of vacillating; a wavering; a moving one way and the other," and the Merriam-Webster dictionary defines vacillate as "to waver in mind, will, or feeling: hesitate in choice of opinions or courses." I might have referred to this aspect as the "FlipFlop" effect or the "U-turn" conundrum. Here are two explanations from Wikipedia and Scienceabc, respectively.

"Loss of multicellularity occurred in some groups. Fungi are predominantly multicellular, though early diverging lineages are largely unicellular (e.g., Microsporidia), and there have been numerous reversions to unicellularity across fungi (e.g., Saccharomycotina, Cryptococcus, and other yeasts). It may also have occurred in some red algae (e.g., Porphyridium), but it is possible that they are primitively unicellular. Loss of multicellularity is also considered probable in some green algae (e.g., Chlorella Vulgaris and some Ulvophyceae). In other groups, generally parasites, a reduction of multicellularity occurred, in number or types of cells". [104]

This "shift" between unicellular and multicellular seems quite simple and straightforward, which begs the question of whether the process is easily reversible? In terms of that first question, whether the process is reversible, it's hard to ignore the research pointing towards <u>vacillation</u> between unicellularity and multicellularity, depending on environmental conditions and the needs of the species. [105]

That's right, folks, you read the excerpts from those two articles correctly.

According to evolutionists, in the oceans of the early Earth, as evolution is at work, some multicellular life forms are apparently not wanting to play evolution's game. Some multicellular life forms have somehow reversed evolution and "evolved" back to unicellular life forms. I am at a loss for words here, but sure, why not. Remember, multicellular life is not a colony of individual cells. Now, remember, earlier, I quoted articles stating that single-celled organisms can, if separated, survive on their own, while cells from a multicellular organism (e.g., liver cells) cannot. And now, evolutionists are saying it is not only possible for multicellular

---

[104] https://en.wikipedia.org/wiki/Multicellular_organism
[105] https://www.scienceabc.com/pure-sciences/how-long-did-it-take-for-multicellular-life-to-evolve-from-unicellular-life.html

organisms to separate, but when they do, those cells can survive on their own as single-celled organisms now.

Granted, these articles speak of this reversal back to unicellular life in the early stages of primitive life, giving examples of fungi and a possibility that it occurred in red algae. Regardless of (according to evolutionists) how primitive the life form is, multicellular organisms carry out their life processes through division of labor by individual cells whose shape and makeup are specific to the function with which that cell performs. These individual cells, if separated, cannot survive on their own, not even in fungi.

Since evolutionists suggest that this "loss of multicellularity" happened in fungi, let's take a quick look at the cellular organization of fungi. "Fungi are eukaryotes and have a complex cellular organization. As eukaryotes, fungal cells contain a membrane-bound nucleus where the DNA is wrapped around histone proteins...Fungal cells also contain mitochondria and a complex system of internal membranes, including the endoplasmic reticulum and Golgi apparatus. Unlike plant cells, fungal cells do not have chloroplasts or chlorophyll. Many fungi display bright colors arising from other cellular pigments, ranging from red to green to black. Pigments in fungi are associated with the cell wall. They play a protective role against ultraviolet radiation and can be toxic. The rigid layers of fungal cell walls contain complex polysaccharides called chitin and glucans. Chitin gives structural strength to the cell walls of fungi. The wall protects the cell from desiccation and predators. Fungi have plasma membranes similar to other eukaryotes, except that the structure is stabilized by ergosterol: a steroid molecule that replaces the cholesterol found in animal cell membranes. Fungi Cell Structure and Function. (2021, March 6). [106]

Wow! So even multicellular fungi, a primitive life form according to evolutionists, is "a complex cellular organization" with cells that perform specific functions and only perform those specific functions with the DNA to prove it. Now that is very compelling scientific evidence for the argument that life was created by a Creator and very damming for the theory of evolution.

In order for fungi to experience a "loss of multicellularity" and evolve backward to unicellular, all those individual cells that have been created and properly arranged to perform their specific function would now have to "evolve" back to a new shape. Also, the DNA that is specific to each type of cell and controls the specific functions of each type of cell would have to be deconstructed and then reconstructed into the original DNA of the unicellular organism. This must happen for the cell to once again have the ability to perform all the required functions of life. And what would happen to some of the organelles within the cell that are no longer needed?

Evolution (evolution is the change in the characteristics of a species over several generations and relies on the process of natural selection) and natural selection (the process whereby organisms better adapted to their environment tend to survive and produce more offspring) move life forward.

Evolutionists cannot tell us how individual single-cell organisms evolved into multicellular organisms or how individual cells in this new life form were able to evolve into cells with unique shapes and functions to make multicellular life possible, but now claim those same cells pulled a U-turn on evolution and evolved back down the evolutionary tree of

---

[106] https://bio.libretexts.org/@go/page/13598

life? I do not even need to ask any common-sense questions here because the evolutionists are at a loss for any explanation.

Even with this "Loss of Multicellularity" that is said to have happened among some groups, the rest of life is moving forward with this new multicellular life, and this evolutionary achievement has only taken 2.9 billion years. Not 2.9 billion years from the formation of the Earth, nope, but 2.9 billion years after the creation of the first life form. 2.9 billion years of evolution created fungi. Outstanding!

Wait, that cannot be right. Could there be debates on the evolutionary timeline of when the first multicellular life came into being? Well, I am sure if there are any differences, they will be minor. Here are four articles representing the range of dates that scientists claim the first multicellular life appeared on Earth.

[107] suggests that the first multicellular life developed around nine hundred million years ago.

[108] claims the first multicellular animals appeared about six hundred million years ago.

[109] claims that it is widely accepted that multicellular life was thought to have begun 575 million years ago. But now, this date has been smashed by the discovery of 2.1-billion-year-old multicellular fossils.

"As mentioned, cyanobacteria may have developed multicellularity quite early—3.5 billion years ago—but the earliest multicellular fungi examples are from 2.5 billion years ago, the oldest plant-like fossils date to about 1.6 billion years ago, the earliest animal fossils appeared around 558 million

---

[107] https://www.newscientist.com/article/dn17453-timeline-the-evolution-of-life/
[108] https://astrobiology.nasa.gov/news/how-did-multicellular-life-evolve/
[109] https://www.iflscience.com/plants-and-animals/oldest-multicellular-life-revealed-detail/

years ago, and multicellular plants evolved from algae around 470 million years ago."[110]

Seriously? That means scientists have pinpointed the timeline for the evolution of multicellular life to between 3.5 billion years and 575 million years ago. That is a range of just over 2.9 billion years!! Isn't the timeline of evolution extremely important to the credibility of evolution? It is extremely frustrating, I know. Now, think of this. Evolution states that the first life evolved around 3.5 billion years ago. The scientists behind the scienceabc.com article claim that multicellular life among some life forms may have evolved 3.5 billion years ago. If that is true, the argument for evolution is pretty much destroyed, and this discovery would be extremely strong evidence that life is the result of a Creator who created all life.

Putting these crazy timelines aside, how do evolutionists propose that life evolved from unicellular to multicellular? Here is an article discussing three theories on the topic to answer that question. To the article's credit, objections to each theory are noted, which are underlined for easy reference.

1) "Symbiotic Theory: This theory suggests that the first multicellular organisms occurred from symbiosis (cooperation) of different species of single-celled organisms, each with different roles. Over time these organisms would become so dependent on each other that they would not be able to survive independently, eventually leading to their genomes being incorporated into one multicellular organism. Each respective organism would become a separate lineage of differentiated cells within the newly created

---

[110] https://www.scienceabc.com/pure-sciences/how-long-did-it-take-for-multicellular-life-to-evolve-from-unicellular-life.html

species. However, the problem with this theory is that it is still not known how each organism's DNA could be incorporated into one single genome to constitute them as a single species."

Excellent question. Look back at everything we have learned about DNA, and then ask yourself this same question. The answer is that it cannot. Next theory.

2) "Cellularisation (Syncytial) Theory: This theory states that a single unicellular organism could have developed internal membrane partitions around each of its nuclei. However, the simple presence of multiple nuclei is not enough to support the theory. To be deemed valid, this theory needs a demonstrable example and mechanism of generation of a multicellular organism from a pre-existing syncytium."

This article actually gives examples of organisms with multiple nuclei and how those organisms do not support this theory. That is very refreshing to see.

3) "The Colonial Theory: The theory claims that the symbiosis of many organisms of the same species (unlike the symbiotic theory, which suggests the symbiosis of different species) led to a multicellular organism. It can often be hard to tell, however, what is a colonial protist and what is a multicellular organism in its own right, as the two concepts are usually indistinguishable, and this problem plagues most hypotheses of how multicellularisation could have occurred." [111]

Out of the hundreds of articles I have read and researched, this is maybe only the second article that states the theory (as a theory, not as fact) and then presents some of the obstacles to that theory. This allows us, the readers, to evaluate the supporting research to the theory and then allows us to form our own opinion on the theory. Heck, who knows, if more

---

[111] https://www.bionity.com/en/encyclopedia/Evolution_of_multicellularity.html

research were presented honestly in this manner, it might inspire the next Albert Einstein, one of the greatest physicists of all time, or Rosalind Franklin, whose work was central to the understanding of the molecular structure of DNA and RNA. These individuals and many other great scientists were free to keep asking the questions of how and why. If someone would have said it just happened and did not allow them to ask questions, we might still be trying to discover the fire.

Evolutionists will state they are not sure about the how and why in the article but then will go on stating as fact the very how and why that they are unsure about. Here is a classic example in this article.

"The different ways that multicellular life developed, however, is *an ongoing subject of debate*... The incredible diversity of form and function that we see on Earth should come as no surprise. *Every form of life on this planet arose from a universal common ancestor* billions of years ago, a single-celled organism with its own unique genetic code that could convert DNA to RNA to proteins. From such humble unicellular beginnings, the entirety of multicellular life on this planet evolved."[112]

How it happened is an ongoing debate, but evolutionists claim that it's a fact that it happened. Using this reasoning, I can claim that the Great Pyramids were built by beavers. I do not know how or why beavers built the Great Pyramids, but I know it is a fact that beavers built the Great Pyramids. Seems about right. Also, from this same article, how does that first single-celled organism with its own genetic code decide to evolve? How does this organism change its genetic code? Answer: It does not decide to evolve, nor does it change its

---

[112] https://www.scienceabc.com/pure-sciences/how-long-did-it-take-for-multicellular-life-to-evolve-from-unicellular-life.html

genetic code. This is not a result of evolution but a Creator who created all life to include single-cell and multicellular organisms.

Now think about this.

We have seen the many articles with different dates on when multicellular life evolved on Earth. Throwing out the extremes, the dates range from 543 million years ago to nine hundred million years ago. In favor of the evolutionists, I am going to use the older date that multicellular life evolved, nine hundred million years ago, as in the new scientist article, to make my point. If life is said to have evolved around 3.6 billion years ago and multicellular life evolved around nine hundred million years ago, that means it took evolution approximately 2.7 billion years to create simple multicellular life in the forms of fungi and algae. That is a long time. Another way to look at this is that evolution took 2.7 billion years to create fungi and algae. Evolution has evolved every other living life form in the following nine hundred million years, including you and me. Do not forget all the life forms that evolved and have gone extinct, for example, the dinosaurs. 2.7 billion years for fungi and algae, under nine hundred million years for everything else. Where is the common sense in that?

Chapter 12

Eons

Geologists generally agree that there are two major eons: the Precambrian eon and the Phanerozoic eon. The Precambrian goes from the formation of the Earth to the time when multicellular organisms first appeared - that's a really long time - from 4,500 million years ago to just about 543 million years ago. Then begins the Phanerozoic eon, which continues up to today."[113]

The Phanerozoic Eon is divided into three Eras: The Paleozoic Era, The Mesozoic Era, and The Cenozoic Era. Each of these "Eras" is broken down into Periods. From the Paleozoic Era, we will be discussing the evolutionary steps from the Devonian and Permian Periods.

The Devonian Period: 419 to 359 million years ago/ Sharks became common at this time. The oldest preserved insects and centipedes appeared as well as the first trees and the forests were present. [114]

The Devonian period lasted sixty million years. For argument's sake, let's pick the middle of the period at 389 million years ago. That would mean evolution evolved life from a "Sponge" (evolved eight hundred million years ago) with no brain, neurons, organs, or tissues to many, many life forms including trees, forests, insects, centipedes, and sharks

---

[113] https://www.geologyin.com/2014/12/geologic-time-scale-major-eons-eras.html
[114] https://www.usgs.gov/youth-and-education-in-science/introduction

in just 411 million years. That is unbelievable! I mean, seriously, that is truly unbelievable. I will not even ask common-sense questions about how aquatic plant life physically moved onto land and evolved into terrestrial plant life, trees, and forest. Also, I am not going to worry about forests, insects, and centipedes. However, we are going to consider sharks. From a sponge to a shark in a mere 411 million years. How advanced and complex are sharks in general without focusing on unique traits that differentiate sharks into five hundred distinct species. Sharks have brains, eyes, hearts, gills, livers, stomachs, tissues, cartilage, blood, and obviously teeth.

Sharks lose teeth during the violent attack on their prey. Sharks bite down on their prey with extreme force. The shark then thrashes, tears, and rips off chunks of its prey. In this process, many teeth are lost. Generations of sharks must have been dying when they could no longer attack prey and feed themselves. According to evolution and natural selection, during this time, some sharks were able to replace teeth, which allowed them to live longer and reproduce and pass this trait on. My common-sense question is if this is how it happened, did evolution try 3 sets of teeth or 10 sets and found that still was not enough? When did evolution determine that to ensure the survival of the shark, it had to grow new teeth continuously?

"A set of new teeth is always developing in the predators' jaw, and they rotate forward like a conveyer belt." [115]

Sharks also have a "Lateral Line." This lateral line, part of the shark's highly developed nervous system, runs lengthwise from the head to the tail down both sides of the shark's body. Sensory cells in this lateral line can detect sound, vibrations,

---

[115] https://www.scientificamerican.com/gallery/sharks-never-run-out-of-teeth/

and pressure changes that alert the shark to the possibility of prey nearby. Do not forget what we just learned about the cells of multicellular organisms. The cells of these organisms are unique in shape and function. The muscle cells of the heart are completely different from the nervous system cells, etc.

Now, ponder the claims of evolutionists that it took 2.9 billion years for evolution to create fungi, algae, and sponges. Then in less than 411 million years, evolution created insects, centipedes, trees, full forests, and the shallow tropical seas. These seas had abundant reefs and a myriad of other sea life, to include the sea life, sharks, that we just learned about. Just try and grasp the diversity and complexity of life that has exploded on the scene in less than 411 million years. Now realize that all these new diverse and complex life forms evolved in 86% less time than it took evolution to evolve one primitive life form, fungi. This means evolution has evolved brains, eyes, hearts, all the organs of these sea life forms, blood, circulatory systems, nervous systems, and digestive systems in this short of a time. Not to mention, evolution has created the male and female sex and sexual reproduction and everything associated with that process. All this evolution and natural selection had taken place in less than 411 million years when it took 2.9 billion years to produce multicellular fungi. I realize that I have repeated this a few times, but that is because it is critically important to see what evolutionists are claiming as fact and demand that we accept it as such. That is, accept it as fact, no questions asked. But common sense tells you that what they are selling just is not right.

But wait, it gets better.

The Permian Period: 299 to 252 million years ago. Reptiles diversified and spread across the land. The *largest mass extinction* of life on our planet occurred at the end of the Permian when ~ 96% of all species perished. [116]

Permian Period (299 to 252 million years ago). The Permian extinction was characterized by *the elimination of over 95 percent of marine and 70 percent of terrestrial species*. In addition, *over half of all taxonomic families present at the time disappeared*. This event ranks first in the severity of the five major extinction episodes that span geologic time. [117]

According to evolutionists, "Therapsids" (which includes mammals) and "Archosaurs" (birds and crocodilians are the only living representatives) survived the Permian extinction. These two groups were able to survive this mass extinction, but how many of them survived? Does it not seem like evolution has to basically start over with 95% of marine life and 70 % of terrestrial life wiped out. Remember, this only happened 252 million years ago. And evolution needs to work fast because the next period has started.

As the Permian Period ends, the Mesozoic Era begins and includes the Triassic, Jurassic, and Cretaceous Periods. The Triassic Period: 252 to 201 million years ago/ After the great extinction at the end of the Permian, many new kinds of animals evolved during the Triassic.

The dominant land animals were reptiles. *The first dinosaurs, marine reptiles, lizards, and tortoises appeared. Crocodiles were abundant.* [118]

See, evolution does not waste any time. After nearly all of life (96% of all species) on Earth was destroyed, dinosaurs and many other species seem to evolve out of the ashes, so to speak. And this new round of evolution happens

---

[116] https://www.usgs.gov/youth-and-education-in-science/paleozoic
[117] https://www.britannica.com/science/Permian-extinction
[118] https://www.usgs.gov/youth-and-education-in-science/introduction

EXTREMELY fast. The Permian Extinction event happened at the end of the Permian Period (299 to 252 million years ago), and the Triassic Period began 252 million years ago.

"Dinosaurs were a successful group of animals that emerged between 240 million and 230 million years ago and came to rule the world until about 66 million years ago. During that time, dinosaurs evolved from a group of mostly dog- and horse-size creatures into the most enormous beasts that ever existed on land". [119]

How can evolution evolve dinosaurs from the surviving species of the Permian Period in only 12 million years? And why is it that the more complex the evolutionary step is, the less time it takes? 96% of all species are wiped out of existence, and in the blink of an eye, evolution evolved the first dinosaurs, marine reptiles, lizards, tortoises, and crocodiles. This just is not passing the common-sense test. How do all this diversity and increasing complexity happen faster and faster? The process of evolution takes place over several generations. Evolutionists, I am sure, will use the argument that the evolution of dinosaurs occurred over the 174 million years that they roamed on planet Earth (240 – 66 = 174). They will say that the dinosaurs did not appear on day one as the Tyrannosaurus rex (T-Rex) or the Supersaurus, measuring between 128-131 feet long, the longest dinosaur to have ever lived. Super, but how do evolutionists expect anyone to believe that? According to the livescience.com article, dog and horse-sized dinosaurs evolved into the Tyrannosaurus rex and Supersaurus in under 174 million years. I say under because there was not just one or two of these monsters roaming around. The evolutionists tell us there were many of these giants on the planet. If so, then this means

---

[119] https://www.livescience.com/3945-history-dinosaurs.html

that these enormous dinosaurs had been around and multiplying for a while now before yet another mass extinction event that, this time, wiped out the dinosaurs. Wow, common sense makes one ask, just how fast did evolution work then, to bring dinosaurs from dog-like creatures to 131-foot long Supersauruses? [120]

If evolution and natural selection are moving life forward for organisms better suited to survive and reproduce in their environment, (Evolution (evolution is the change in the characteristics of a species over several generations and relies on the process of natural selection) and natural selection (the process whereby organisms better adapted to their environment tend to survive and produce more offspring) move life forward.) How do evolutionists explain multiple new life forms evolving at the same time? Especially if that species or many species have been nearly wiped out due to a mass extinction event. For instance, how do evolutionists justify saying evolution evolved dinosaurs, tortoises, and crocodiles in the same environment at the same time? How does that line up with the definition of evolution and natural selection?

With the end of the Triassic Period, The Jurassic Period (201 to 145 million years ago) begins and ushers in a time of giant plant-eating dinosaurs roaming the Earth with smaller but vicious carnivores stalking them. Flying reptiles and the first birds appeared, and creeping about in the undergrowth were tiny mammals no bigger than rats. If the first dinosaurs were the size of dogs and horses, why did they evolve to such monstrous sizes? The answers given by evolutionists are just plain silly. Anyone with any common sense knows that it

---

[120] https://en.wikipedia.org/wiki/Supersaurus

cannot happen the way they say. Here are three articles that prove my point.

"Today's largest land animals are the elephant and the giraffe, and both became large as a result of different advantages their size gave them. The gigantism of the giraffe seems obvious: its long neck allows it to reach for higher foliage that is inaccessible to other herbivores, giving it an advantage in gathering food. To support such a long neck, a larger body is also needed. Large animals are also more difficult to hunt. If one elephant is large and another is small, predators will have an easier time catching and killing the smaller ones. Thus, the bigger elephant survives and has offspring, which may be slightly larger than its parents in adulthood. Over time, elephants reach their present-day size. . . . Sauropods are a unique group of dinosaurs. They had hollow bones, didn't chew their food, had incredibly long necks, and likely possessed huge stomachs. These traits are theorized to be key in how they attained their enormous size. Undoubtedly, their long necks allowed them to reach food other animals couldn't, which made a bigger size more advantageous for them. Their long necks relied on two key traits: hollow, or pneumatized, bones of the spine, as well a small head, which allowed the neck to be light."[121]

"Why did some dinosaurs grow so big? Paleontologists don't know for certain."[122]

"Dinosaurs lived during the Triassic, Jurassic, and Cretaceous. The climate was much warmer during these periods, with CO_2 levels over four times higher than today. This produced abundant plant life, and herbivorous dinosaurs may have evolved large bodies partly because there was

---

[121] https://www.scienceworld.ca/stories/how-did-dinosaurs-get-so-big/
[122] https://www.usgs.gov/faqs/why-did-some-dinosaurs-grow-so-big

enough food to support them. But being large also helps to protect against predators. The giant sauropods had to eat plants as fast as they could to grow big enough to be safe from carnivores like T. rex and Spinosaurus. Meanwhile, the carnivores were becoming larger just so they could tackle their enormous prey. Another possibility is that the herbivorous dinosaurs were ectothermic (cold-blooded), and being huge helped them regulate their temperature.

This theory is problematic though because evidence increasingly suggests that the large carnivores were endothermic (warm-blooded)." [123]

Now for a few common-sense thoughts and questions on what these articles are claiming. Also, by now, do you see how these articles are written so that you, the reader, walk away believing the proposed suggestion is actually a fact by the way the article is written? That is dishonest, to say the least. Are we supposed to actually believe that the giraffe evolved from an original species that were supposedly about the size of deer, and because they kept reaching up with their necks to eat, evolution changed the entire structure of the animal to create the giraffe? Evolution evolves the giraffes, yes, giraffes, to feed where other animals cannot, but now because of this evolutionary change, giraffes are more vulnerable to predators. Is that not a big oops for evolution? I guess not even evolution can get it perfect. Also, why did only one species evolve into giraffes when all the deer-like original species exhibited that same feeding behavior? And just to make your head spin a little bit more, here is an article that contradicts the earlier article on the theory of why giraffes have long necks. "There is little evidence that feeding is the reason the

---

[123] https://www.sciencefocus.com/nature/why-were-dinosaurs-so-big/

long neck evolved, and the current theory is that it helps males compete for females during the mating season."[124]

I am not even surprised anymore when article after article disagrees or contradicts other articles' claims with regard to evolution. I am surprised that evolution is still the law of the land; no questions asked.

Now for the elephant example, if evolution created larger elephants because predators killed the smaller elephants more than the bigger elephants, how does evolution justify that the current size of elephants still makes them vulnerable to predators? If evolution allowed them to get a little bigger, is it possible that the only predator that could hunt them would eventually be humans? And on that train of thought, why did evolution not evolve the elephants bigger or faster to counter ancient man, who hunted them with spears?

The reason for the Sauropods growing so big is a splendid example of, "you gotta be kidding me!" The article does not flat out say it, but it seems like it is suggesting that the sauropods evolved two key traits: hollow, or pneumatized, bones of the spine, as well a small head, which allowed the neck to be light. These key traits allowed the neck to get longer to reach food sources that other dinosaurs could not. The other article says that sauropods and other dinosaurs, prey, and predators reached such large sizes due to all the available food. Think of one of many modern-day restaurants and the all-you-can-eat buffet. And the dinosaurs did not even have to worry about getting a clean plate on each return trip to the buffet.

The first question about the Sauropods is, did evolution hollow out normal neck bones? Also, did evolution shrink its

---

[124] https://sciprogress.com/giraffe-vs-human-neck-bones/

head? Now I suggest you read this article, [125] especially these two sections, "Air-sac system" and "Architecture of Sauropod Necks." From advanced air-sac systems that included cervical air-sacs and extensive cervical diverticula running the full length of the neck, to having the perfect "air space proportion (ASP) of a bone is the proportion of its volume taken up by pneumatic cavities" is further evidence that a Creator and not evolution designed life. Another amazing point is that this article and many others do an amazing job at detailing the supposed results of evolution in these magnificent creatures but do not address the process of evolution to evolve the new life form. Common sense asks how evolution evolved these pneumatic cavities and every other evolutionary change needed for this species to live?

And as for the theory that the dinosaurs were eating as much and as fast as they could due to the abundant food source, which kicked evolution into gear, allowing them to evolve larger and larger, that is false. This theory has been proven false by every animal on planet Earth, including humans. If a species just eats and eats, it only gets fatter and fatter. The most honest answer to the question of why some dinosaurs grew so large is from the USGS article. "Paleontologists don't know for certain." That is a perfectly good and honest answer. A refreshing answer. The truth is that dinosaurs, elephants, and giraffes are the size they are because the Creator created them that way. And just like that, we are beginning a new period.

The Jurassic Period ends as the Cretaceous Period (145 to 66 million years ago) begins. One evolutionary creation that I really find interesting is that grasses supposedly evolved during this period. I am very intrigued by this because,

---

[125] https://www.ncbi.nlm.nih.gov/pmc/articles/PMC3628838/.

according to the evolutionary timeline, as far back as The Permian Period (299 to 252 million years ago), temperate forests were abundant with "conifers." Some examples of conifers are Cedars, Cypresses, Douglas-firs, Firs, Junipers, Redwoods, Spruces, and Pines. During The Jurassic Period (201 to 145 million years ago), conifers continued to be the most diverse large trees, and Cycads (evergreen, cone-bearing, palm-like plants) became so abundant and diverse that the Jurassic is often called the "Age of Cycads," but grasses had not yet come to be. Evolution figured out how to move plant life from the oceans to land and create the giant redwood trees but had not created grasses yet. That is just odd when trying to follow how evolution through natural selection moves life forward. What is the purpose of grasses, then?

Anyway, the Cretaceous Period has begun, and it is a joyful time as the dinosaurs and marine reptiles are flourishing. Also, during this period, many new species are appearing. On the avian side of life, the birds are diversifying, and their populations are growing and flourishing as well. There are many new mammals that evolved during the Cretaceous period, including the three groups of mammals that live today. Those groups are the Monotremes, which lay eggs. If you are stuck on what mammals lay eggs, do not worry. It is a very small group. Only 5 species of mammals lay eggs, one of which is that odd-looking Duck-Billed Platypus. The second group is The Marsupials which keep the newborn in a pouch. There are over 250 species of marsupials. Probably the most well-known marsupial is the Kangaroo, and a lesser-known marsupial is the Tasmanian devil. Did you just get a flashback of the Bug's Bunny cartoons on Saturday morning, featuring the Tasmanian devil, like I just did? Ok, I might have just dated myself a little bit. Moving on to the third and final group are the Eutherian or 'Placental' mammals. This group is characterized by the presence of a placenta, which is a vascular organ that facilitates the exchange of nutrients and waste between the mother and the fetus. Humans are among this group, along with between 4,000 and 5,000 other species.

I am sure at this point; common sense is having you ask these questions: Why are all these species diversifying, and so many new species being created? If evolution is diversifying and creating new species better suited to the environment, why are the old species surviving? Why are both species not only surviving but flourishing? According to evolution's own definition, it should not be happening. And if this has made you wonder, or you have thought of a few questions of your own, wait till you read the next paragraph.

"One of the most significant developments during the Cretaceous was the appearance and rapid diversification of the first flowering plants . . . Due to the appearance of flowering plants, many modern groups of insects appeared and began to diversify, including ants, termites, bees, butterflies, aphids, and grasshoppers." [126]

Are evolutionists claiming that evolution created these insects because of the flowering plants? By how the article was written, yes, it does appear that is what is being suggested, but I cannot honestly believe that is what they are claiming. Although, according to the evolution timeline of these insects (three articles noted below), it is possible that some evolutionists are absolutely suggesting that.

"Ants first appeared on the earth between 140 to 168 million years ago." [127]

---

[126] https://www.usgs.gov/youth-and-education-in-science/mesozoic
[127] https://expeditions.fieldmuseum.org/australian-ants/ant-evolution-and-environment

"Bees evolved from ancient predatory wasps that lived 120 million years ago."[128]

"Grasshoppers diverged in the mid to late Cenozoic Era (~65 million years ago to the present)"[129]

I think I am just going to leave it at that and let you decide for yourself what you think is being suggested.

Another significant event of the Cretaceous Period was when the oceans were starved of oxygen and could have been the reason for another mass extinction of marine organisms. See below.

"Oceanic anoxic events have been recognized primarily from the already warm Cretaceous and Jurassic Periods when numerous examples have been documented. Oceanic anoxic events have had many important consequences. It is believed that they have been responsible for mass extinctions of marine organisms."[130]

Just as the oceans were recovering from the mass extinction of marine life due to a lack of oxygen, the Mesozoic Era ended with a bang, quite literally. An exceptionally large meteorite slammed into Earth with extreme force. The site of the impact is buried underneath the Yucatan Peninsula in Mexico, near Chicxulub Puerto and Chicxulub Pueblo, and why the impact site is named the Chicxulub crater. The meteorite is estimated to be just over six miles in diameter. This impact caused massive destruction through heat waves and tsunamis. This event was referred to as the K-T extinction, which is an abbreviation of Cretaceous-Tertiary extinction but is now referred to as the K-Pg extinction event.

According to many evolutionists, the K-Pg extinction is what killed off *all* the dinosaurs, yet many other species were

---

[128] https://www.museumoftheearth.org/bees/evolution-fossil-record
[129] http://entomologytoday.org
[130] https://en.wikipedia.org/wiki/Anoxic_event

able to survive. Some articles claim that the K-Pg extinction wiped out up to 80% of all animal and plant species. [131]

Seems to be a little bit of a disagreement on the impact "pun intended" of this event. Just the dinosaurs or was it a much larger extinction, and you might be asking yourself, like I have many times, if this event was global, how did every dinosaur die, but other species survive? Why did avian (flying) dinosaurs die, but birds survived? Here are two articles offering possible explanations.

Russ Graham, senior research associate in geosciences at Penn State. "It was the huge amount of thermal heat released by the meteor strike that was the main cause of the K/T extinction," Graham explains, adding that underground burrows and aquatic environments protected small mammals from the brief but drastic rise in temperature. In contrast, the larger dinosaurs would have been completely exposed, and vast numbers would have been instantly burned to death. After several days of searing heat, the Earth's surface temperature returned to bearable levels, and the mammals emerged from their burrows, Graham notes. But it was a barren wasteland they encountered, one that presented yet another set of daunting conditions to be overcome. According to Graham, it was their diet that enabled these mammals to survive in habitats nearly devoid of plant life. "Because most of the earth's above-ground plant material had been destroyed." Mammals, in contrast, could eat insects and aquatic plants, which were relatively abundant after the meteor strike. As the remaining dinosaurs died off, mammals began to flourish. Although representatives from other classes of animals also

---

[131] https://www.usgs.gov/youth-and-education-in-science/mesozoic

survived the K/T extinction—crocodiles, for instance, had the saving ability to take to water."[132]

This theory is so vague and assumes the reader cannot think on their own. First, apparently, only mammals that could burrow and marine life in the oceans were able to survive. Second, the Earth's temperature rose so fast that any life form not in a burrow or water would burn to death, but in only a couple of days, the Earth's temperature was bearable enough for the burrowed mammals to explore this new devastated and barren environment. If this was the case, did any of the oceans, lakes, or other bodies of water instantly evaporate, killing the life within? The conditions were so horrific and hot that the planet was devoid of nearly all plant life, yet insects survived. How? Did all the mammals burrow? Were there carnivorous mammals, and if so, what did they eat during the K-Pg extinction event? And last but not least, why did the small dinosaurs, the same or smaller in size to the mammals, go extinct but not the mammals?

As described in the articles below, mammals were more than little mouse-like creatures hiding during the dinosaur era. So again, why did only the dinosaurs go extinct globally when birds and mammals survived?

"Repenomamus is a genus of an opossum- to badger-sized…mammal, containing two species, Repenomamus robustus, and Repenomamus giganticus. Both species are known from fossils found in China that date to the early Cretaceous period, about 125-123.2 million years ago. In fact, Repenomamus was larger than several small sympatric dromaeosaurid dinosaurs like Graciliraptor R. robustus discovered with the fragmentary skeleton of a juvenile Psittacosaurus (small dinosaur) preserved in its stomach

---

[132] https://www.psu.edu/news/research/story/probing-question-why-did-mammals-survive-k-t-extinction/

represents the second direct evidence that at least some Mesozoic mammals were carnivorous and fed on other vertebrates, including dinosaurs." [133]

((According to the myth, a world crowded with dinosaurs left little room for mammaliaforms (ancient mammals and their ancestors). As a result, mammals and their kin remained tiny, mouse-like, and primitive. The myth posits that mammals didn't evolve diverse shapes, diets, behaviors, and ecological roles until the K-Pg mass extinction event 66 million years ago killed off the dinosaurs and "freed up" space for mammals. "This is a very old idea, which makes it very hard to defeat," said David Grossnickle, a postdoctoral researcher in the Department of Biology at the University of Washington. "But this view of mammaliaforms simply doesn't stand up to what we and others have found recently in the fossil record." New species arose that, for example, could climb, glide or burrow — and ate more specialized diets of meat, leaves, or shellfish.)) [134]

This paper by Richard Cowen uses K-T instead of K-Pg and discusses multiple theories on the K-T extinction event. Very honest pros and cons to each theory discussed, with a lot of possible evidence to support or counter the theories. Here is the link. [135]

One frustrating part of the research for this book has been trying to find a straight answer to many questions, for example, what species went extinct and what survived this extinction event. Thankfully, in his paper's first two paragraphs, he clearly states what went extinct and what survived this event. What a welcomed change.

---

[133] https://en.wikipedia.org/wiki/Repenomamus
[134] https://www.washington.edu/news/2019/06/20/mammaliaforms-ecological-radiation/
[135] https://ucmp.berkeley.edu/education/events/cowen1b.html

"Almost all the large vertebrates on Earth, on land, at sea, and in the air (all dinosaurs, plesiosaurs, mosasaurs, and pterosaurs) suddenly became extinct about 65 Ma, at the end of the Cretaceous Period. At the same time, most plankton and many tropical invertebrates, especially reef-dwellers, became extinct, and many land plants were severely affected. This extinction event marks a major boundary in Earth's history, the K-T or Cretaceous-Tertiary boundary, and the end of the Mesozoic Era. The K-T extinctions were worldwide, affecting all the major continents and oceans. There are still arguments about just how short the event was. It was certainly sudden in geological terms and may have been catastrophic by anyone's standards."

"Despite the scale of the extinctions, we must not be trapped into thinking that the K-T boundary marked a disaster for all living things. Most groups of organisms survived. Insects, mammals, birds, and flowering plants on land, and fishes, corals, and mollusks in the ocean went on to diversify tremendously soon after the end of the Cretaceous. The K-T casualties included most of the large creatures of the time, but also some of the smallest, in particular the plankton that generates most of the primary production in the oceans." He goes on to remind the reader that if "any of the theories that try to explain only the extinction of the dinosaurs ignore the fact that extinctions took place in land, sea, and aerial faunas, and were truly worldwide." A very important fact that many theories overlook. Cowen goes on to point out the mystery of why birds survived this event, and avian dinosaurs did not. "The survival of birds is the strangest of all the K-T boundary events if we are to accept the catastrophic scenarios. Smaller dinosaurs overlapped with larger birds in size and in ecological roles as terrestrial bipeds. How did birds survive while dinosaurs did not? Birds seek food in the open by sight; they are small and warm-blooded, with high metabolic rates and small energy stores. Even a sudden storm or a slightly severe winter can cause high mortality among bird populations. Yet an impact scenario, according to its enthusiasts, includes "a nightmare of environmental disasters, including storms, tsunamis, cold and darkness, greenhouse warming, acid rains, and global fires." There must be some explanation for the survival of birds, turtles, and crocodiles through any catastrophe of this scale, or else the catastrophe models are wrong."

Richard Cowen sums up this paper by stating some of the unknowns that still exist. This is important because it informs future scientists on where to focus their research rather than attempting to build up research based on total fabrication.

"We still do not have an explanation for the demise of the victims of the K-T extinction while so many other groups survived. We do not know whether it was the impact alone or the combination of the impact and the plume volcanism that caused the extinction, and we do not know the linkages between the physical events and the biological and ecological effects. It would be astonishing if the impact played no role, and it would be astonishing if the volcanism played no role." [136]

How insulting to our intelligence is that first article. The meteorite burnt the dinosaurs, and then the mammals took over and flourished. How awesome is the second article? It clearly states which species went extinct and which did not and then investigated some of the theories objectively. I like how he ended the paper by stating some of the unknowns and

---

[136] https://ucmp.berkeley.edu/education/events/cowen3b.html

stating some of his personal thoughts as such and not as facts, by saying "he would be astonished if."

It is now approximately 66 million years ago, and life has again been hit with two mass extinctions during the Cretaceous period, one in the oceans with a loss of oxygen and the second being the K-Pg extinction event wiping out 80% of all animal and plant species. This is where evolutionists talk about all the new life forms that evolved now that the dinosaurs are extinct. Again, I ask, why does evolution need to evolve anything? The mammals that were prey to the dinosaurs have hit utopia. Why is it necessary to evolve, and why is it necessary to evolve into entirely new life forms? The K-Pg extinction event is the true end to the Cretaceous Period and the start of the Cenozoic Era.

Cenozoic Era (66 million years ago until today) brings forth many new life forms in the animal and plant kingdoms. Evolution works very quickly to overcome the last two mass extinctions during the Cretaceous Period. Now referring back to the definition of evolution and natural selection, (evolution (evolution is the change in the characteristics of a species over several generations and relies on the process of natural selection) and natural selection (the process whereby organisms better adapted to their environment tend to survive and produce more offspring) move life forward.) Common sense has some questions. Armadillos, primitive primates, and the first rodents are some of the first new creations of evolution during the Cenozoic Era. How are these newly evolved animals better adapted to the environment than their predecessors?

Early in the Cenozoic Era was also when cacti and palm trees appeared. "Recent molecular phylogenetic work has confirmed that Pereskia, a genus that consists of 17 species of leafy shrubs and trees, is where the earliest cactus lineages began."[137] Common sense begs to ask the question, why? According to evolutionists and the definition of evolution, change takes place over generations. Yet, common sense tells us that if any of these seventeen species of shrubs and trees are somehow planted in a desert-type environment, they are going to die off because they are not cacti and not suited for life in that environment. Why would evolution continue to throw these shrubs and trees into this harsh environment? And this article claims that some of the traits needed to survive in these conditions were developed before the split to cacti; common sense asks why? What is the point of developing traits that are not yet needed? Does evolution know the future?

WHALES

Whales came to be during the Eocene Epoch (56 – 34 million years ago), and if you still do not have any doubts about evolution up to this point, I assure you that the story of how whales evolved will make you question evolution. According to the evolutionary timeline, life began in the oceans of early Earth, and some of these aquatic life forms evolved into terrestrial life forms and roamed the land. Then approximately fifty million years ago, one or more of these terrestrial mammals ditched land life and evolved back into an aquatic life form. Sure, at this point, why not, and if it sounds a little far-fetched, here is one of many articles claiming just that.

"Both hippos and whales evolved from four-legged, even-toed, hoofed (ungulate) ancestors that lived on land about 50 million years ago. Modern-day ungulates include hippopotamus, giraffe, deer, pig, and cow." "The theory is that

---

[137] https://www.sciencedaily.com/releases/2006/05/060515000551.htm

some land-living ungulates favored munching on plants at the water's edge, which allowed them to hide from danger in shallow water easily. Over time their descendants spent more and more time in the water, and their bodies became adapted for swimming. Their front legs became flippers, and a thick layer of fat called blubber replaced their fur coats to keep them warm and streamlined. Eventually, their tails became bigger and stronger for powerful swimming, and their back legs shrunk. Gradually, their nostrils moved to the top of their heads so that they could breathe easily without the need to tilt their heads while swimming. As some of these creatures began to feed on a different diet, they evolved into baleen filter feeders and lost their teeth."[138]

This seems more like a story from the imagination of a six-year-old or the basic premise for a new Steven Spielberg movie. First off, the article claims that the land-living ungulates could hide from danger in shallow water. Sharks and crocodiles are prominent during this time, and both of these groups hunt in shallow water. That would be like the old saying, "out of the frying pan and into the pot." I do not see the evolutionary advantage. Next, the ungulates' front legs became flippers because they spent more and more time in the water. For argument's sake, let us go along with this line of thinking. At some stage of evolution, before the ungulate spent all of its life in the water, the front legs would be useless for walking and useless as a flipper. How would these ungulates at this stage of the evolutionary change be able to mate and reproduce to continue passing along this trait of front leg to flipper, which at this point would just be called useless front "thingy."

---

[138] https://us.whales.org/whales-dolphins/how-did-whales-evolve/

The next step that evolution took replaced the fur with a thick layer of fat, which on whales is called blubber. Well, Polar bears have both fur and blubber, and they also spend a lot of time in the water. So how did evolution know to ditch the fur and beef up the blubber on the ungulates but not the polar bears? Now the tail gets stronger and stronger and even changes shape in order to become strong swimmers. Change shape, you say? The article did not mention changing shape. Correct. Think of any animal you wish. Now, common sense asks, how many of these tails are shaped like a whale's tail? The answer is none. Evolution not only has to strengthen and elongate the whale's tail, but it also has to change the shape and purpose of the tail completely. Ever wonder how many different whale tails did evolution try before choosing the present shape? Or did evolution get lucky and create the perfect tail on the first try? Is there fossil evidence of evolution trying different tail shapes?

Apparently, because whales are swimming all the time now, the back legs just shrank up into the body. Why did evolution create the dorsal fin for stabilization and not evolve the back legs into "pelvic" fins like those of a shark, which are used for stabilization? Yes, a shark also has a dorsal fin, but why would evolution not use the back legs instead of just letting the back legs go to waste? And how did the dorsal fin evolve? Evolutionists also claim that the nostrils gradually move to the top of the head to assist in breathing air while swimming. This is so plainly stated that one would feel foolish even to ask the simplest question about it. The migration of the nostrils to the top of the whale's head is stated in the same manner as when marine biologists talk about the migration of whales from the cold-water feeding grounds to the warmer waters for mating. Can it really be that simple?

Here is an extremely basic and brief description of the breathing process of a whale.

"When the animal inhales, air passes through the blowhole, nasal duct, larynx, trachea, and lastly the lungs. . . During evolution, blowholes migrated to the top of the head, which facilitates breathing at the water surface. Whales cannot breathe through their mouth because, unlike terrestrial mammals, their digestive system and respiratory system are not connected. The blowhole leads to the nasopharynx or nasal duct. . . The higher pressure experienced during a dive compresses the alveoli. Air is then pushed toward the bronchioles and bronchi, which do not collapse under pressure due to their cartilaginous wall."[139]

Again, this article claims as fact that the nostrils migrated to the top of the whale's head during evolution and details the pathway air travels to reach the whale's lungs. The article also states that whales cannot breathe through their mouths due to a separation of the digestive and respiratory systems, unlike terrestrial mammals. Now I can see you are asking yourself but wait, don't evolutionists claim that a terrestrial mammal that breathed through its mouth evolved into a whale? Yes, that is what evolution claims. This means that the digestive and respiratory systems had to totally separate at some point, and in the respiratory system, evolution had to develop the cartilaginous walls of the bronchi so they would not collapse under the higher pressures exerted during deep dives. All of this is ignored and stated simply as the nostrils moved to the top of the head.

Remember the definition of evolution. Evolution (evolution is the change in the characteristics of a species over several generations and relies on the process of natural

---

[139] https://baleinesendirect.org/en/discover/life-of whales/physiology/respiratory-system/

selection) and natural selection (the process whereby organisms better adapted to their environment tend to survive and produce more offspring) move life forward. Now if evolution is true and terrestrial mammals did evolve into whales, can you even fathom how many generations and how many failed changes it would have taken evolution to actually create a whale?

Let common sense remind us that the DNA sequence in the genetic code would have to be changed to reshape the skull. This new DNA sequence would have to be altered each and every time the nostrils supposedly moved along this migration to the top of the head. The DNA coding that determines the positioning of the skin, muscles, blubber, and nerves would also have to be re-written each and every time there was a change in the position of the nostrils. The size of the nostrils must also change as the whale increases in size, as well as the function of the nostrils. The nostrils of the terrestrial mammal do not have to physically close, but the nostrils of the whale must not only close but must also be watertight when closed. The nostrils of the whale must also be able to stay watertight with the ever-increasing pressure asserted against the nostrils as the whale dives deeper and deeper. Do not forget that all these physical changes require changes in the brain to control this new breathing technique and all the other changes occurring. And this is barely scratching the surface of all the changes that would actually have had to take place for a four-legged, even-toed, hoofed (ungulate) to evolve into a whale.

How many incomplete changes in "characteristics" were made along the evolutionary journey to create a whale? Think about it, if this is really how whales came to be, along the evolutionary timeline of whales, you would have whales with partial front limbs/ flippers and shrinking back legs. Tails would grow thicker and develop a weird, flattened shape at the end of the tail, all while trying to figure out how to eat as its digestive and respiratory systems are separating. And hopefully, at each step along this evolutionary change, the nasal duct and larynx are continuing to line up with the holes in the skull, as are the physical nostrils as it all migrates from the front of the face to the top of its head. Holy smokes, is that a lot to assume just happens. And that is only mentioning the major components that would have to change. Now try to comprehend all the changes that must be made on a cellular level to achieve these changes. This is not an insult to your intelligence but an honest question. Can any of us truly comprehend what must happen on a cellular and genetic code level to achieve any of this? Now, on top of all that, these ungulates were herbivores at the beginning of this evolutionary journey, and now they are carnivorous. That is no small feat in and of itself. Although according to evolutionists, ungulates to whales is settled science. I can assure you that it is not settled and that whales did not evolve from four-legged, even-toed, hoofed (ungulate) mammals. A Creator created whales.

With the evolution of whales fresh in our minds, let me ask this. In the Oligocene Epoch (34 – 23 million years ago), many species perished because they could not survive the changes in climate that were happening, and many new forms evolved that could handle these changes. Common sense asks another simple question. Why did evolution have to evolve new life that could survive these changes in climate? If evolution can change a terrestrial herbivore into a carnivorous whale, would not evolution be capable of evolving the existing species to survive the changes in climate instead of evolving entirely new life forms capable of surviving the new climates?

How about this interesting claim about the evolutionary change in horses during the Oligocene Epoch. Horses increased in size, developed longer and stronger legs, and reduced their toes to hooves to achieve the ability to run faster. How did evolution figure out this combination? How many trials and errors were there to get to horses as we know them?

I hope you are beginning to realize what a leap of, dare I say it, *faith* it takes to accept evolution.

Then in the Miocene Epoch (23 – 5 million years ago), hoofed animals with multiple stomachs, which were needed to digest the tough grasses properly, flourished along with the increasing grasslands. Horses and cows can graze on the same grasses. That being a fact, common sense asks why evolution would see the need to create multiple stomachs in cows to digest grasses but only one stomach to digest the same grasses in horses? Developing multiple stomachs seems like a lot of unnecessary evolution instead of just improving the digestive ability of one stomach. And from our discussion on whales, imagine what must happen on a molecular level to evolve additional stomachs. Physical changes must also be made to accommodate the space that these additional stomachs need, just to name a few.

Now in the Pliocene Epoch (5 – 2.6 million years ago), evolutionists claim many animals become extinct because of new competition. Let common sense ask this if evolution can evolve longer necks for giraffes and the Supersaurus to overcome competition for food, how or why does evolution not intervene here and evolve these animals to compete better? Doesn't this seem to be in direct opposition to evolution and natural selection? Shouldn't some traits have evolved that better suited those animals and their environment, and those traits would then be passed on through the generations?

During the Pliocene Epoch is when evolutionists claim that apes start the evolution into early hominids. And with that, we come to the beginning of the evolutionary tale of humans, you and me. From a single cell formed by random chance in the primordial soup to the apes that will evolve into our ancestors. (Here is your clue, Evolution. Alex, what is a fictional story claimed as fact by evolutionists.).

Chapter 13

Genus; Homo

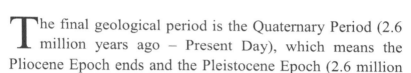

The final geological period is the Quaternary Period (2.6 million years ago – Present Day), which means the Pliocene Epoch ends and the Pleistocene Epoch (2.6 million years ago – 10,000 years ago) begins.

"The first humans emerged in Africa around two million years ago, long before the modern humans known as Homo sapiens appeared on the same continent."[140]

What are the defining characteristics of the genus "Homo"? What differentiates us from other species? Would you believe that there is much debate over those exact questions? Does that surprise you? Hopefully not at this point. At least the evolutionists are consistent. Scientists, evolutionists, or whatever label they carry do not agree on any single evolutionary milestone other than to demand that evolution is fact and the only possible explanation for the origin of and progression of life from the very beginning to the present day.

With regards to the human species, many scientists disagree about the line from which humans evolved and the traits and characteristics that separate the genus Homo from Australopithecus. The two most agreed-upon characteristics that have resulted in us having a separate genus are bipedalism (walking on two legs) and our higher intellect or intelligence.

---

[140] https://www.history.com/news/humans-evolution-neanderthals-denisovans

Equipped with that knowledge, let us see where evolutionists claim we evolved from.

"Humans (genus Homo) may have descended from australopith ancestors, and the genera Ardipithecus, Orrorin, Sahelanthropus, and Graecopithecus are the possible ancestors of the australopiths. Most scientists maintain that the genus Homo emerged in Africa within the Australopiths around two million years ago. *However, there is no consensus on within which species.*"[141]

"Various Eurasian and African Miocene primates have been advocated as possible ancestors to the early hominins, which came on the scene during the Pliocene Epoch (5.3–2.6 mya). *Though there is no consensus among experts. . .* That we and the extinct hominins are somehow related and that we and the apes, both living and extinct, are also somehow related is accepted by anthropologists and biologists everywhere. Yet the exact nature of our evolutionary relationships has been the subject of debate."[142]

How can Britannica claim this relationship is accepted by anthropologists and biologists everywhere, but there is debate on the nature of the relationship? This relationship is obviously not accepted everywhere since anthropologists and biologists are still debating it. Shame on Britannica for intentionally trying to deceive and mislead its readers. People use sources like Britannica to try and find accurate and factual information. Opinion cannot be presented as fact. That is wrong no matter what.

Regardless of one's belief in life being the result of evolution or life being created by a Creator, facts and opinions need to be presented as such. How does one trust anything else

---

[141] https://en.wikipedia.org/wiki/Australopithecine
[142] https://www.britannica.com/science/human-evolution

from sites, books, papers, or research articles when it is discovered that opinion is being passed off as fact? Continuing on now with what classifies the human mammal as a human.

"Human evolution is characterized by a number of morphological, developmental, physiological, and behavioral changes that have taken place since the split between the last common ancestor of humans and chimpanzees. *The relationship between all these changes is the subject of ongoing debate.*" [143]

"Most scientists currently recognize some 15 to 20 different species of early humans. However, s*cientists do not all agree about how these species are related or which ones simply died out.* Many early human species – certainly the majority of them – left no living descendants. *Scientists also debate how to identify and classify particular species of early humans and what factors influenced each species' evolution and extinction.*" [144]

"The genus Homo consists of eight species, including our own. The species from the oldest to the youngest with regards to when they lived are; H. Habilis (2.4 – 1.4 mya), H. Rudolfensis (1.9 – 1.8 mya), H. Erectus (1.89 mya – 110,000 ya), H. Heidelbergensis (700,000 – 200,000 ya), H. Neanderthalensis (400,000 – 40,000 ya) also called Neanderthals, H. Naled (335,000 – 236,000 ya), H. Sapiens (300,000 – present) we are homo-sapiens, H. Floresienis (100,000 – 50,000 ya)." [145]

Common sense would like to ask the question, why are Homo-sapiens the only species of Homo still alive? That is an excellent question, especially when you consider this; "We

---

[143] https://en.wikipedia.org/wiki/Human
[144] https://naturalhistory.si.edu/education/teaching-resources/social-studies/human-evolution
[145] https://humanorigins.si.edu/evidence/human-fossils/species

found 6,495 species of currently recognized mammals (96 recently extinct, 6,399 extant)."[146]

And that is just mammals. In 2011 calcademy.org reported a new study that estimated the total number of species on Earth. The estimate reported 6.5 million species found on land and 2.2 million dwelling in the ocean depths. That is a total of 8.7 million species (the article says give or take 1.3 million).[147]

It is incredible that every life form on planet Earth has many, many varieties of that species. Species that supposedly evolved, lived, and competed for food and mates alongside each other in the same geographical areas. Not only did these life forms compete for food with other species of its own genus, but species of other genuses as well, and all of the individual species have survived and flourished. Look at Killer Whales, Dolphins, Porpoises, Sharks, and other fish-eating whales, which have flourished even though these species can occupy the same geographical waters while hunting for prey.

Every life form, every species has managed to survive alongside other species of its genus, all but one species that is, and that would be humans. Evolutionists claim that humans are the only species of its kind. This both helps and hurts the argument for evolution. Without other living species of humans, it helps the evolutionary argument that when fossils are discovered that are similar (sometimes it is a real stretch to say that the fossils are similar) to the human skeleton structure, evolutionists classify the fossil as an ancestral or closely related species to Homo sapiens. The flip side to this argument is that out of all life forms on planet Earth with

---

[146] www.researchgate.net/publication/322962382_How_many_species_of_mammals_are_there
[147] www.calacademy.ort/explore-science/how-many-species-on-earth

many, many related species, Homo sapien is the only species with zero related surviving species. How is this statistically possible? It is not. This next article tries to make the claim that Homo sapiens evolved and out-competed our ancestors, eventually driving them to extinction.

"This process of gradual mutation takes a lot of time as species reproduce. When subject to the same environmental pressures, the species will either compete and become dominant or die out from lack of survival now that they're being out-competed. This process takes hundreds of thousands of years. Finally, adaptation is where the majority of a species has now transitioned into the newer form. Its genetic material has gradually altered with each generation and is now the predominant expression of that species on Earth. As we'll see with human evolution, it's then common for the old variants to die out and the new species to continue with success." [148]

Why is it that evolution only causes the old variants of the human species to "die out" and not the old variants of every other life form since bacteria? One argument is that Homo sapiens were too intellectually advanced. This higher intellect allowed Homo sapiens to develop more advanced tools and weapons, which were then used to out-compete our ancestors. Could that be true? Think of the world back then. There was more than enough land and food to support all the human species. Even if Homo sapiens were out-competing all the other species of Homo, why wouldn't evolution evolve the other species to be able to compete with the Homo sapiens or the species that were being out-competed migrate to another location instead of going extinct? Think of the world today; there are still several Indigenous tribes in Africa, South

---

[148] https://www.mybiosource.com/learn/complete-guide-to-human-evolution/

America, and other countries that still live like the early years of Homo sapiens. How do those tribes live alongside our technology without going extinct? Although they are also Homo sapiens, evolutionists claim that our advanced technology of tools and spears drove the early humans to extinction. I would say our technology has advanced a bit since then, from hunter-gatherers to agriculture to having food delivered using an app on a smartphone. Common sense would like to ask, why hasn't this technology driven these tribes to extinction? If these tribes can still exist alongside modern man, then surely our early ancestors could also. And now that Homo sapiens is the king of the jungle, so to speak, another debate among scientists arises. Scientists cannot agree on whether evolution has stopped or if modern humans are actually influencing evolution.

Some paleontologists believe that human evolution has slowed down or even stopped like Stephen J. Gould. He posited that evolution operates by punctuated equilibrium, which just means that the Darwinian evolutionary process happens in short bursts and then plateaus for a period of stability, where no noticeable changes occur. Like any scientific theory, we don't know for certain if that is the case. Of course, the claim that evolution has slowed or stopped has its challenges. American geneticist Alan R. Templeton argued against Attenborough's theory by asserting that our cultural and technological developments happen alongside and even inform our environment, so if anything, it'd help natural selection. In case you couldn't tell yet, the answer is a big "maybe."

"As time marches on, there's no doubt that these competing theories will be proven right or wrong, or maybe replaced by new ones entirely, kind of like evolution itself." [149]

Common sense reminds us that evolution cannot pause or stop by the very definition of evolution. Evolution (evolution is the change in the characteristics of a species over several generations and relies on the process of natural selection) and natural selection (the process whereby organisms better adapted to their environment tend to survive and produce more offspring) move life forward.

Bipedalism – walking on two legs

Bipedalism, a defining trait of the genus Homo and modern-day humans, Homo sapiens. The evolutionists have proposed some very unique theories on why humans might have developed this trait of bipedalism. Theories that make me wonder if they are just seeing how silly of a theory they can put forth, and it still is accepted by the scientific community and the general public? Theories like these two articles.

"The initial changes toward an upright posture were probably related more to standing, reaching, and squatting than to extended periods of walking and running." [150]

I hope this first article gave you a good laugh. Britannica actually published an article stating that humans evolved bipedalism because we were standing and reaching more. I mean, come on, this is just flat-out silliness. How does anyone still lend any credibility to the theory of evolution with theories like this? Honestly. If you told someone that early humans developed bipedalism because they started standing

---
[149] https://www.mybiosource.com/learn/complete-guide-to-human-evolution/
[150] https://www.britannica.com/science/human-evolution/Background-and-beginnings-in-the-Miocene

and reaching more, they would tell you to stop wasting their time. Common sense asks the question, why would we be standing more if our skeletal frame was not set up to allow longer periods of standing? Standing would be physically uncomfortable and possibly painful, so what would make us want to stand more? Also, what were we reaching for that the other primates were not reaching for? Wouldn't it seem more likely that if we were in the same geographical locations that we would all be reaching for the same objects? Primates are excellent tree climbers and have longer arms, so what was out of reach? This next article suggests two theories. Let me warn you though, for the first theory, if you are a romantic, you might want to go grab some tissues before reading it.

"Anthropologist C. Owen Lovejoy of Kent State University revived Darwin's explanation by tying bipedalism to the origin of monogamy. . . As climatic changes made African forests more seasonal and variable environments, it would have become harder and more time-consuming for individuals to find food. This would have been especially difficult for females raising offspring. At this point, Lovejoy suggests, a mutually beneficial arrangement evolved: Males gathered food for females and their young, and in return, females mated exclusively with their providers. To be successful providers, males needed their arms and hands free to carry food, and thus bipedalism evolved. Another theory considers the efficiency of upright walking. In the 1980s, Peter Rodman and Henry McHenry, both at the University of California, Davis, suggested that hominids evolved to walk upright in response to climate change. As forests shrank, hominid ancestors found themselves descending from the trees to walk across stretches of grassland that separated forest patches.

Rodman and McHenry argued that the most energetically efficient way to walk on the ground was bipedal. [151]

Ok, right off the top of my head, if this theory is correct, then only the male's developed bipedalism. How did the females develop this ability? Although, my favorite part of this article is the claim that the male developed bipedalism because his arms were carrying food. I guess evolution had no choice but to evolve an entirely new species that could walk upright just so that this new species could carry food in both its arms. Now my common-sense question here is, why make the arms shorter on this new "human" species? Wouldn't the longer arms allow this new species to carry more food? Carrying more food per trip would save energy. Just a thought.

According to evolutionists, the male species has an evolutionary need to mate with as many females as possible to ensure the survival of its DNA. If this is true, then what benefit does monogamy provide the male? A single mating partner does not guarantee the survival of the males' DNA. The sole purpose of mating for the male is to pass along his DNA. As stated in the following article, males will perform horrendous acts to ensure their DNA is the only DNA to be passed along.

"Male animals of many other species are known to kill young infants that aren't related to them. This allows them to mate with the mother and to have more of their own offspring in the group." [152]

If our closest ancestor is the chimpanzees which are not monogamous, why would we develop a monogamous mating relationship?

---

[151] https://www.smithsonianmag.com/science-nature/becoming-human-the-evolution-of-walking-upright-13837658/
[152] https://www.newscientist.com/article/2150258-male-chimpanzee-seen-snatching-seconds-old-chimp-and-eating-it/

"Chimpanzees, which have a promiscuous mating system."[153]

The last theory in the smithsonianmag.com article suggests that bipedalism evolved from having to walk across stretches of grasslands due to shrinking forests. Then that would mean, if chimpanzees had to move across stretches of grasslands, they should have also developed bipedalism, correct?

"Movement over any significant distance usually takes place on the ground. Though able to walk upright, chimpanzees more often move about on all fours, leaning forward on the knuckles of their hands (knuckle-walking)."[154]

The evolution of humans' distinct bipedal gait remains a focus of research and *debate*. [155]

Regardless of the "why" evolutionists think that humans developed bipedalism, the physical differences between humans and chimpanzees that evolution would have had to overcome are astounding. Yet, evolutionists state it as a plain fact that these changes are part of our evolutionary past. As you read on, remember our discussion about the evolution of the whale's nostrils and the actual changes that would have had to happen, from DNA to the physical changes to the whale's skull. That was just moving two holes in the skull.

Now in the case of walking upright because of evolution, the new species' entire skeleton has to be changed. From the shape of the skull so it can now balance on top of the newly designed spine to the redesigned feet and everything in between, including the nostrils. Here is an article to highlight some of the changes that evolution would have had to make.

"Perhaps the most important difference is that chimps cannot extend their knees and lock their legs straight as humans can. Instead, they have to use muscle power to support

---

[153] https://en.wikipedia.org/wiki/Monogamy_in_animals
[154] https://www.britannica.com/animal/chimpanzee
[155] https://www.pnas.org/doi/10.1073/pnas.1715120115

their body weight when standing or walking upright, a much more tiring situation. In humans, the thigh bone slopes inward from the hip to the knee, placing our feet under our center of gravity. We also have well-developed muscles (called gluteal abductors) on the side of our hips that contract to prevent our bodies toppling to one side when all our weight is on one foot in mid-stride. Chimps have thigh bones that do not slope inward to the knee like ours, so they stand and walk with their feet wide apart. Their gluteal abductors are also much weaker than ours, so they have to rock their whole body from side to side during each step in order to move their center of gravity over whichever leg is bearing their weight. Human bodies have a number of other adaptations to walking upright, as well. Our foot is specialized as a weight-bearing platform, with an arch that acts as a shock absorber. Our spines have a characteristic double curve, which brings our head and torso into a vertical line above our feet. The surfaces of the joints in our legs and between our vertebrae are enlarged, which is an advantage for bearing weight. And the hole through which the spinal cord enters the skull, called the foramen magnum, is near the center of the cranium in humans, allowing our heads to balance easily atop our spines rather than toward the back of the cranium as in chimps."[156]

Amazing how perfect we seem to have evolved. Think of all the DNA and cellular changes any one of these changes would entail. It is truly overwhelming trying to think of it all, so let's start with a smaller change. Chimpanzees and humans both have the same basic hand structure with four fingers and a thumb. Has the evolution of the human hand helped or hindered our genus, Homo? With that question, let's find out.

---

[156] https://www.pbs.org/wgbh/evolution/library/07/1/l_071_02.html

Opposable Thumbs

"Strong and nimble thumbs meant that they could better create and wield tools, stones, and bones for killing large animals for food. Because developing dexterous, opposable thumbs pushed our ancestors to make and use tools, eat more meat and grow bigger brains, scientists have long wondered if such thumbs began only with our own genus, Homo, or among some earlier species. Shorter thumbs and longer fingers are helpful for climbing. But as our ancestors forsook life in the trees and increasingly began to make and manipulate objects, shorter fingers and longer opposable thumbs would have produced a hand assembly that got better and better at grasping. Over time, natural selection could have refined these anatomical changes based on the many ways humans used their hands and which of those proved most rewarding, like smashing animal bones to collect their high energy marrow." [157]

I bet vegetarians do not like this article stating that we only grew bigger brains and became smarter because our ancestors ate more meat. I wonder if there are any vegetarian evolutionists and what their thoughts are? But I digress. Common sense asks, how are better tools built if opposable thumbs have not evolved yet? Did our now smarter brain want to build better tools but could not because of our thumbs? Was it the sheer desire to build those better tools that caused evolution to evolve opposable thumbs on our ancestors? Now, if the other primates saw our ancestors being more successful hunting with tools, wouldn't you think they would try it and evolve alongside our ancestors? That is just common sense.

---

[157] https://www.smithsonianmag.com/science-nature/how-dexterous-thumbs-may-have-helped-shape-evolution-two-million-years-ago-180976870/

Why would they watch us succeed while they struggled for survival? Although, if our ancestors and the other primates were doing just fine, why would the genus Homo need to be evolved?

"It has been proposed that the hominid lineage began when a group of chimpanzee-like apes began to throw rocks and swing clubs at adversaries and that this behavior yielded reproductive advantages for millions of years, driving natural selection for improved throwing and clubbing prowess. This assertion leads to the prediction that the human hand should be adapted for throwing and clubbing. . . The typical primate hand is characterized by a diminutive thumb in combination with long, curved fingers (Midlo, 1934). In contrast, the human hand has a much larger, more muscular, mobile, and fully opposable thumb combined with fingers that have shortened and straightened. This striking exception to the primate pattern clearly requires an evolutionary explanation (Marzke & Marzke, 2000; Fig. 1)."[158]

I like this mainly because of the sentence, "This striking exception to the primate pattern clearly requires an evolutionary explanation." Yes, clearly, this has to be evolution and not a different species that was created by a Creator. Putting that aside, why did only our ancestors develop the opposable thumb alongside other primates that did not? And are we to believe that because our ancestors kept having difficulties swinging clubs or throwing rocks, that evolution stepped in and, over time, shrunk our four fingers and grew our thumb longer? How many molecular changes had to happen to achieve this? Think about the muscular and tendon make-up of our hands, and what changes would be required to achieve the functionality of the human hand?

---

[158] https://www.ncbi.nlm.nih.gov/pmc/articles/PMC1571064/

Besides opposable thumbs, primates are covered with thick fur, and we are not. Why? There are three main theories on this conundrum.

Loss of Fur

(A more widely accepted theory is that when human ancestors moved from the cool shady forests into the savannah, they developed a new method of thermoregulation. Losing all that fur made it possible for hominins to hunt during the day in the hot grasslands without overheating. An increase in sweat glands, many more than other primates, also kept early humans on the cool side. Sarah Millar, a co-senior author of the new study and a dermatology professor at the University of Pennsylvania's Perelman School of Medicine, explains that *scientists are largely at a loss to explain why different hair patterns appear across human bodies.* "We have really long hair on our scalps and short hair in other regions, and we're hairless on our palms and the underside of our wrists and the soles of our feet," she says. *"No one understands really at all how these differences arise.")* [159]

"Eccrine sweat glands—millions of which produce our salty, cooling perspiration— and chimpanzees do have eccrine pores across their bodies. Chimpanzees, one of our closest primate relatives, with whom we share nearly 99% of our genome, rely heavily on panting in hot weather." [160]

If chimps have eccrine sweat glands as we do but pant to cool down, as stated in the sciencefriday article, why would evolution evolve an entirely new system to regulate genus Homo's body temperature? Now the Smithsonian mag article

---

[159] https://www.smithsonianmag.com/science-nature/why-did-humans-evolve-lose-fur-180970980/
[160] https://www.sciencefriday.com/articles/sweat-glands-evolution/

states that evolution shed the fur and increased the number of sweat glands when our ancestors moved onto the savannah. Common sense asks were there no chimps on the savannah, and if there were, why did evolution not shed their fur? Also, the primates' fur is basically one length, so why would humans develop different types of hair on different parts of our bodies. The hair on our heads grows very long. The hair on our arms and legs are short. Men have more hair than females. Another question is, why did the fur of the primates become less coarse to become the hair of humans? Scientists have no idea why.

"By losing our fur, the human body was able to sweat a lot more, allowing our body to cool down a lot quicker. This theory is also used to describe why we stopped walking on four limbs; as the ground retained heat, keeping the body closer to the ground would have meant that the human body would cool down much slower. Another perspective of this theory is also that our ancestors starting to walk on only two limbs required more effort from the body, therefore, our body was heating up a lot more than it was used to."[161]

This theory is great because both points contradict evolution. The first point is that walking on four limbs kept the body close to the retained heat in the ground, increasing the time needed to cool down. If this is true, common-sense asks, why haven't all the other primates evolved to at least walking upright to achieve the ability to cool down faster? The second point claims the body was heating up more due to the increased effort required from walking upright. Every other evolutionary article published on walking upright claims it to be a much more efficient method of travel with less energy being exerted. That means the body would not have heated up more, and theoretically, our ancestors could have kept the fur.

---

[161] https://historyofyesterday.com/why-humans-lost-most-of-their-body-hair-171d79eee2d6

"As bipeds, or animals that walk upright on two legs, our heads are directly exposed to the sun. Near the equator, where humans evolved, sun exposure can be overbearing, and head hair helps people avoid overheating. "It's sort of a built-in hat," Pagel said. Head hair also helps retain heat at night. "Our brains are relatively small compared to the rest of our bodies, but they're enormously metabolically active," Pagel said. This activity produces heat, and head hair could insulate this area of concentrated warmth." [162]

What about the rest of our bodies? Our heads were protected from the sun with our new fancy "built-in hat," but the rest of our bodies would have fried! Our skin would have been white, just like the skin of chimpanzees underneath their fur, as stated in the following article.

Chimpanzees are covered by a coat of brown or black hair, but their faces are bare except for a short white beard. Skin color is generally white except for the face, hands, and feet, which are black. [163]

This white skin would not have done well in the "overbearing" blazing sun near the equator where we supposedly evolved. Humans would have gone extinct from skin cancer before we learned how to spell melanoma.

Look at all the unknowns and discrepancies among the scientists on just the four topics in this chapter on human evolution. I did not even deep dive into any of these four topics, and still, there are a plethora of articles making a claim followed by a statement that the claim is in debate. Imagine the craziness I would find if I deep-dived into any of these subjects or other evolutionary milestones like the evolution of the eye or hearing. So many unsubstantiated claims about

---

[162] https://www.livescience.com/why-hair-on-head.html
[163] https://www.britannica.com/animal/chimpanzee

evolution. Yet we are told it is a fact, and questioning evolution is ludicrous.

With that, I would like to move to the final chapter, where I will remind you of a few actual facts, observations, and points that I have made along the way, as well as ask a few final questions.

CHAPTER 14

Evolution Defeated

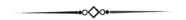

I would like to give you a little bit of perspective by starting this final chapter with what is printed on my bag of almonds.

"Scientific evidence suggests, but does not prove, that eating 1.5 ounces per day of most nuts, such as almonds, as part of a diet low in saturated fat and cholesterol may reduce the risk of heart disease."

Now think about that for just a minute in comparison to the claims evolutionists make about the creation of life. If scientists cannot figure out if nuts can reduce the risk of heart disease, how are we supposed to accept, without question, the outrageous claims evolutionists and scientists make about evolution.

Congratulations. Give yourself a pat on the back because we did it. Wow. I hope this has been as eye-opening and thought-provoking of a journey for you as it has been for me. Writing this book has been the realization of a calling that I have had in my heart for years now, to show that evolution just does not hold up to simple, common-sense questions. We started at the beginning of the universe, the singularity, and finished with the evolution of the genus Homo. Taking on the foundational theories in support of evolution and natural selection armed only with everyday common sense, those foundational theories crumbled one after the other, giving clear and convincing evidence of the existence of a Supreme Being, the Creator, also known as God.

Since we did cover so much, 4.6 billion years' worth of evolution, to be exact, I did want to end the book with a cliff notes version, so to speak. Some of the bigger points of disagreement that I feel show evolution is not the answer to life or the progression of life on planet Earth.

We started some 4.6 billion years ago to the very instant that scientists say the universe began. That moment is defined as the singularity. A point of zero volume and infinite density. From the singularity, the ever-expanding and unimaginable vastness of the universe was born. A universe that contains an estimated 100 billion to 200 billion galaxies, with each of these galaxies, depending on their size, containing "just a few hundred million stars to giant galaxies with one hundred trillion stars," according to Wikipedia. In this universe, we find our own galaxy, the Milky Way. The Milky Way Galaxy created our solar system and our planet Earth. All this creation, according to evolutionists, is the result of random chance. On our very own planet, Earth, random chance brought forth the first protein. After random chance created this first protein, evolutionists credit evolution and natural selection for the development and advances of all further life forms, from the first bacteria to you and I here today, including all life that has gone extinct.

Although, scientists have no answer or even really any solid theories on what created the singularity. It is kind of hard to say random chance, evolution, and natural selection created the universe, our galaxy, Earth, and all life while ignoring what put it all into motion. Honestly, regardless of one's religious beliefs, the only possible answer to how all the matter of the entire universe was condensed into a point of zero volume, and infinite density and then released all of that energy and matter, setting the universe into motion is without a doubt, God. God is the Creator of all. It is just common sense.

Next, all matter is made up of atoms, which are made up of protons, neutrons, and electrons. All the matter of the universe that was compressed into the singularity is composed of different combinations of these three subatomic particles. The universe did not start expanding, and then these subatomic particles formed by random chance. Think about how amazing it is that these building blocks of everything, including the book you are reading and the air that you are breathing, are engineered so perfectly.

All protons are uniform in size and carry a positive electric charge. All neutrons are uniform in size and carry a neutral electric charge, and all the electrons are uniform in size and carry a negative electrical charge. Without this perfect relationship among these three particles, there would be no life. Not only would there be no life, but there would also not be any water, rocks, air, planets, stars, or galaxies. Matter is only possible because of this perfect relationship between these subatomic particles.

After all the "must have happened" and "scientists are not in agreement" statements regarding evolution, can anyone still honestly accept that the absolutely perfect relationship between these three subatomic building blocks of life is due to random chance? No, I didn't think so.

Can anyone still accept that random chance created these building blocks of matter and then compressed all matter into the singularity? Again, no, I didn't think so.

The very existence of the perfect relationship of the proton, neutron, and electron and the fact that *all* matter is made up of different combinations of these three subatomic particles absolutely proves beyond the shadow of a doubt that the universe and everything in it was created by the hand of God.

Ignoring the singularity, which I think is impossible to ignore, could random chance really create the perfect uniformity of the subatomic building blocks with their perfect electric charges? Could random chance really create the first correctly sequenced protein? Remember the conservative odds of that are 10 to 164th. That is one correctly sequenced functional protein chain for every 10 to 164th failed attempts (1 right attempt for every 100 million, trillion, trillion, trillion, trillion, trillion, trillion, trillion, trillion, trillion, trillion, trillion, trillion failed attempts. To see it written out looks like: 1 success in

100,000,000,000,000,000,000,000,000,000,000,000,000,000,000,000,000,000,000,000,000,000,000,000,000,000,000,000,000,000,000,000,000,000,000,000,000,000,000,000,000,000,000,000,000,000,000,000,000,000,000,000,000,000,000,000,000,000,000,000

failed attempts. Even with odds this bad, evolutionists still somehow deny the hand of God in the creation of life.

Evolutionists really do not have any answer from the protein to the formation of a cell, yet simply state that it had to happen. Well, that is obvious because there is no life without the formation of cells. Evolutionists only talk of the formation of a cell in small steps, which has been disproven. Remember Ann Gauger, (Developmental Biologist) at the Biologic Institute, who said, "You can't do it one bit at a time because everything works together in a causal loop. The higher level of organization transcends the pieces. The spatial organization in the cell requires that molecules end up in the right place at the right time." Paul Nelson, (Philosopher of Biology) at Biola University said this. "The simplest living cell we know has more than 300 different proteins... carbohydrates, complex sugars, nucleic acids, DNA & RNA, lipids, and a whole variety of different chemicals which jointly constitute the living state." We are supposed to believe that the complexity of this arrangement was the result of random chance. You know in your soul random chance has nothing to do with this, and this is true evidence of the Creator. If that didn't jog your memory as to the blind assumptions that evolution requires, here is another one you will remember.

Cells are either prokaryotic with no nucleus or eukaryotic with a nucleus. And the evolutionist's explanation for how prokaryotic cells evolved into eukaryotic cells was that one cell ingested another cell and, instead of consuming it, created a mutual relationship with the cell. Come on now. Common sense easily shoots this theory down. The sole purpose of that first cell engulfing the other cell is for food. Why would it not consume it? How does it not consume it? How did the engulfed cell communicate to the cell that just engulfed it? Then there are all the questions regarding the DNA of the two cells. Also, how was this evolutionary technique distributed to all the other cells if it did happen? Scientists do not have answers for any of these questions other than to say, "it must have happened." The complexity of cells is further evidence that this was not random chance but could only be accomplished by the hand of God.

Why the need for evolution?

Everything up to this point is the result of random chance, according to evolutionists. Evolution and natural selection didn't enter the picture until there was life, which we now have with a fully functioning cell. Single-celled life is the beginning of all life. Now that we have life, a great common-sense question is, why is there even a need for evolution? Remember that *evolution (evolution is the change in the characteristics of a species over several generations and relies on the process of natural selection) and natural selection (the process whereby organisms better adapted to their environment tend to survive and produce more offspring) move life forward.* At this point, life is single-celled. Bacteria was the first, and it conquered the planet. By the very definition of evolution and natural selection given to us by evolutionists, there is no need for evolution. Bacteria seem to be perfectly adapted to its environment, which we learned was basically all of planet Earth.

"One of the key concepts of evolution is adaptation, closely accompanied by the Darwinian theory of Natural Selection. As environments and other species have changed and developed, individual species had to adapt to these changes. This was greatly attributed to the process of natural selection. Those that were not fit or suited to the environment died off, while those that survived were able to pass on their genetic material. Over generations of repetition, some genes died out completely, while others became dominant. This, in turn, changed the characteristics of species, eventually creating new ones altogether...." [164]

Again, how does the original life form not die off when evolution evolves a new life form or species? Another obvious common-sense question was, why doesn't evolution evolve the entire species? The answer in this quote basically sums up the case against evolution for me.

---

[164] https://northamericannature.com/how-did-whales-evolve/

"The reason other primates aren't evolving into humans is that they're doing just fine," Briana Pobiner, a paleoanthropologist at the Smithsonian Institute in Washington, D.C., told Live Science. [165]

How awesome is this quote? If one species is doing fine, then why does a subset of that species evolve into another species. If the original species is doing fine, there is no need for evolution. If there is no need for evolution, then how does one explain the vast diversification of life on the third rock from the sun, our planet Earth? I think you know my answer. God creates life.

Consider this as well, evolution does not better adapt any species for its environment, nor does evolution solve any problems. Evolution only creates new problems and struggles that the new species must now overcome to ensure its survival. Every time a new life form is "evolved," new problems arise. Two basic but obvious problems are how does this new life form eat and reproduce? I guess we are to assume that for as long as it takes a species to evolve, these changes are occurring uniformly among males and females that are continuing to mate and produce mutated "evolved" offspring. What has evolution solved?

If we try and accept that evolution and natural selection are responsible for all life, then how do evolutionists explain the in-between stages of evolution? Evolution does not happen all at once; it is the result of small genetic changes that are passed from parent to offspring over many generations until the new desired life form is realized. With that in mind, you can really pick any life form to compare the evolution of the old life form to the new life form. But since we have discussed whales earlier in the book, I am going to use them for my comparison.

---

[165] https://www.livescience.com/32503-why-havent-all-primates-evolved-into-humans.html

Now whales were land mammals that supposedly moved back to life in the seas. Evolution claims that the more time these animals spent in the water, their front limbs evolved into flippers. At some point along that evolutionary journey, it would be reasonable to think that the front limbs would not be strong enough to support the animal on land, but these limbs would not be fully developed flippers yet. They would be known as front "thingy's," and who knows what the hind limbs look like at this stage.

This evolution of the species can only happen if the evolving species can find and mate at each stage of evolution. Each species would need to be in the same stage of evolution for the required adaptation of the genetic code needed to further the evolution of the species to be passed along to the offspring. Would there be some point when the species is not capable of mating along this evolutionary journey? I believe there would be. If and when that happens, that species would cease evolving and die off because it could not mate and reproduce. If the species cannot mate, it cannot pass on the new genetic code.

Evolution does not even attempt to answer this question other than to say the fossil record does not have much evidence of species in transition. Common sense gives you the obvious answer that you already know. A species could not change according to how evolution says it changes. This new species is because the Creator created it for a purpose.

Do not forget that each time life seemed to have been cruising along just fine or reached a pinnacle; there was some form of mass extinction that killed off a majority of life and/or the life forms. Evolution must basically start all over to a degree. And with that, is it not astounding that the more complicated the evolutionary step, the shorter amount of time that was needed for evolution to evolve new species. Remember that it took one billion years for evolution to evolve the first cell, using the evolutionary timeline in the newscientist.com article. Then it took evolution roughly 2.9 billion years to evolve from unicellular life, bacteria, to multicellular life, simple sponges. Sponges have no brain, neurons, organs, or tissues. Now apparently, evolution is on fire because, in only 440 million years, life evolved from simple sponges to many aquatic life forms, including sharks. Remember how amazing sharks are, including the highly developed lateral line on sharks. Over the next 405 million years, there were three mass extinction events, including the *"The Permian extinction was characterized by the elimination of over 95 percent of marine and 70 percent of terrestrial species. In addition, over half of all taxonomic families present at the time disappeared. This event ranks first in the severity of the five major extinction episodes that span geologic time."* [166]

Does something as major as mass extinction events bother evolution. Nope, as a matter of fact, somehow, these events seem to speed up evolution. I ask you, where is the common sense in this line of thinking? Does evolution have an office with genetic blueprints stored somewhere so that when there is a mass extinction event, evolution just grabs the blueprints and moves on with evolving the next species? I know I am being silly there, but are not some of evolution's claims just as silly?

Honestly, let's take a step back, take a deep breath and look at what is outside our window. Whether you are looking at a tree growing up through the sidewalks of New York City, lush redwood forest of California, whale watching at the Strait of

---

[166] https://www.britannica.com/science/Permian-extinction

Gibraltar, or studying chimpanzees in the forest north of the Congo River, you can see the hand of God at work. Life is not just an accident that happened by random chance and furthered along by evolution.

It is so amazing that when you look at planet Earth and all the life contained on Earth, every life form is perfectly situated to fill a role and a purpose. Just like the cells of our body with their unique roles and responsibilities, so are the life forms on Earth. How perfect that even the most basic forms of life, bacteria, and fungi, play such an important role in life. They are fundamental in so many ecosystems driving biochemical cycles. They recycle nutrients like nitrogen to the soil for plants to use, which provides food for the animals and humans (you and me). Then there are insects like ants and earthworms that help aerate the soil allowing oxygen and water to reach the roots and cycling the nutrients of the soil and are part of the food chain for other life forms. Sponges filter out bacteria in fresh and saltwater environments and produce a carbon food source for the sea life to feed upon, which promotes another food source for humans. Flowering plants produce nectar which attracts honeybees, and as the honeybees are sucking up the nectar, the flower's pollen sticks to the honeybee. The honeybee is now spreading the pollen as it moves from flower to flower, plant to plant. The nectar is then turned into honey back at the hive. Honey can be used as an anti-inflammatory, antioxidant, and antibacterial agent. Honey has also been used to treat coughs and as a topical treatment for burns and wounds. Honey can also be stored indefinitely. The plants that the honeybees are helping with pollination and the trees of the Earth are taking in carbon dioxide and releasing the oxygen that marine and terrestrial life need to exist.

Everything has been created for a purpose. That purpose was created by God, not by a combination of random chance, evolution, and natural selection. Without God creating us for a purpose and creating everything else, there is no reason for evolution or for life. All life was created for a purpose. God has given you a purpose, and you can feel it in your soul. It might be deep down, and you have not discovered it yet, but it is there. Continue to see the evidence of God in everything and continue to ask God what His purpose is for you. He will answer.

Full Disclosure

My Faith

I am a Christian. I believe the Bible is God's word and that God created the heavens above and the Earth below, including man, woman, and every living creature. I am writing this book because I believe that science proves the existence of God. Just as we have seen in this book. Science has made tremendous discoveries with the chemical make-up, structure, and functions that protein, DNA, and RNA perform, just to name a few. With every new scientific discovery, there is new evidence showing the hand of God in the creation of life and the universe.

Simple, common-sense questions are so important because they should keep the scientific community honest. If scientists cannot answer these questions, then the hypothesis that has been suggested should not be presented as factual. More study, research, and experiments are needed to either prove or disprove the hypothesis. Science welcomes questions and challenges; it is a shame that some scientists do not.

"In fact many scientists do believe in biblical creation, Creation scientists - creation.com [https://creation.com/creation-scientists]" -

Thank you for reading, and never stop asking common-sense questions.

If after reading this book you see God's hand in all of creation, congratulations. God is calling you to be counted among the saved. Do not ignore this calling. The more you ignore His calling, the less you will hear His calling, and one day it might be too late.

Read these scriptures. What is God telling you?

Romans 1:20 (NKJV) For since the creation of the world, His invisible attributes are clearly seen, being understood by the things that are made, even His eternal power and Godhead, so that they are without excuse.

Romans 3:23 (NKJV) for all have sinned and fall short of the glory of God.

John 3:16 (NKJV) "For God so loved the world that He gave His only begotten Son, that whoever believes in Him should not perish but have everlasting life.

Romans 10:9-10 (NKJV) that if you confess with your mouth the Lord Jesus and believe in your heart that God has raised Him from the dead, you will be saved. For with the heart, one believes unto righteousness, and with the mouth, confession is made unto salvation.

John 20:31 (NKJV), but these are written that you may believe that Jesus is the Christ, the Son of God, and that believing you may have life in His name.

If you are ready to be saved and have your name written in the book of life, trust God and invite Him into your heart by praying this prayer.

I confess that I am a sinner and ask for forgiveness of my sins. I believe in and trust Jesus Christ as my Lord and Savior, who died for my sins, rose from the dead, is alive and sitting at the right hand of God.

Acknowledgments

I would like to thank my parents, who were loving, supportive, and amazing role models for me throughout my life. They followed Proverbs 22:6 Train up a child in the way he should go, And when he is old, he will not depart from it. (NKJV).

From my brother Robert to my "passing friends" Pete and Steve, who have always spoken the truth even when it is difficult to hear, I thank you for being there for me.

To my wife and my children, who each in their own special ways, have impacted my life in so many ways and have made me the man that I am today.

Thank you all for being in my life.

SHANE ANDERSON, AUTHOR OF COMMON-SENSE VS. EVOLUTION.

Shane Anderson was born in Palm Beach Gardens, Florida. His early years were spent going on several camping trips with his family in Florida and vacationing in the mountains of North Carolina. Shane is retired military, having served in both the Navy and Army, as well as a graduate of the University of North Carolina, Charlotte. Shane was part of a group of four motorcyclists who rode their bikes straight through from San Diego, CA, to Jacksonville, FL, in only 47 hours. Faith in God is the one absolute truth in life. God loves you no matter what.

Made in the USA
Columbia, SC
24 November 2024